Production Note

Cornell University Library produced this volume
to replace the irreparably deteriorated original. It
was scanned at 600 dots per inch resolution and
compressed prior to storage using CCITT/ITU
Group 4 compression. The digital data were
used to create Cornell's replacement volume on
paper that meets the ANSI Standard Z39.48-
1992. The production of this volume was
supported by the United States Department of
Education, Higher Education Act, Title II-C.

Scanned as part of the A. R. Mann Library
project to preserve and enhance access to the
Core Historical Literature of the Agricultural
Sciences. Titles included in this collection are
listed in the volumes published by the Cornell
University Press in the series *The Literature of
the Agricultural Sciences,* 1991-1996, Wallace
C. Olsen, series editor.

HEREDITY

IN RELATION TO EVOLUTION
AND ANIMAL BREEDING

HEREDITY

IN RELATION TO EVOLUTION
AND ANIMAL BREEDING

BY

WILLIAM E. CASTLE

PROFESSOR OF ZOÖLOGY, HARVARD UNIVERSITY

NEW YORK AND LONDON
D. APPLETON AND COMPANY
1911

Published September, 1911

Printed in the United States of America

PREFACE

THIS little book is based on a course of eight lectures delivered in November and December, 1910, before the Lowell Institute, Boston, as well as on a course of five lectures delivered before the Graduate School of Agriculture held under the auspices of the Association of Agricultural Colleges and Experiment Stations at Ames, Iowa, in July, 1910. The hope is entertained that it may be of service to students and that it will also interest the general reader.

The writer wishes to express his gratitude to the Carnegie Institution of Washington for permission, in its preparation, to draw freely upon published and unpublished material derived from investigations aided by the Institution.

Acknowledgment is also due to the following persons, or to their publishers, for permission to use figures from their publications, as indicated in the text: Prof. E. B. Wilson and The Macmillan Co., Prof. H. S. Jennings and *The American Naturalist*, Dr. W. B. Kirkham and The American Book Co.

W. E. CASTLE

JUNE, 1911

CONTENTS

LIST OF ILLUSTRATIONS

HEREDITY

INTRODUCTION

GENETICS, A NEW SCIENCE

THE theory of organic evolution has probably influenced more fields of human activity and influenced them more profoundly than has any other philosophic deduction of ancient or modern times. By this theory philosophy, religion, and science have been revolutionized, while in the practical arts of education and agriculture, twin foundation stones of the state, man has been forced to adopt new methods of procedure or to justify the old ones in the light of a new principle.

The evolutionary idea has forced man to consider the probable future of his own race on earth and to take measures to control that future, a matter he had previously left largely to fate. With a realization of the fact that or-

ganisms change from age to age and that he
himself is one of these changing organisms man
has attained not only a new ground for humility
of spirit but also a new ground for optimism and
for belief in his own supreme importance, since
the forces which control his destiny have been
placed largely in his own hands.

The existence of civilized man rests ultimately
on his ability to produce from the earth in sufficient abundance cultivated plants and domesticated animals. City populations are apt to
forget this fundamental fact and to regard with
indifference bordering at times on scorn agricultural districts and their workers. But let the
steady stream of supplies coming from the land
to any large city be interrupted for only a few
days by war, floods, a railroad strike, or any
similar occurrence, and this sentiment vanishes
instantly. Man to live must have food, and
food comes chiefly from the land.

A knowledge of how to produce useful animals
and plants is therefore of prime importance.
Civilization had its beginning in the attainment
of such knowledge and is limited by it at the
present day. If, therefore, this knowledge can

be increased, civilization may be advanced in a very direct and practical way. Before Darwin the practices of animal and plant breeders were largely empirical, based on unreasoned past experience, just as was in antiquity the practice of metallurgy. Good plows and good swords were made long before a scientific knowledge of the metals was attained, but without that scientific knowledge the wonderful industrial development of this present age of steel would have been quite impossible. In a similar way, if not in like measure, we may reasonably hope for an advance in the productiveness of animal and plant breeding when the scientific principles which underlie these basic arts are better understood. Two practical problems present themselves to the breeder: (1) how to make best use of existing breeds, and (2) how to create new and improved breeds better adapted to the conditions of present-day agriculture. We shall concern ourselves with the second of these only.

The production of new and improved breeds of animals and plants is historically a matter about which we know scarcely more than about the production of new species in nature. Selection has been undoubtedly the efficient cause of change in both cases, but how and why applied and to what sort of material is as uncertain in one case as in the other. The few great men who have succeeded in producing by their individual efforts a new and more useful type of animal or plant have worked largely by empirical methods. They have produced a desired result but by methods which neither they nor any one else fully understood or could adequately explain. So there exists as yet no true science of breeding but only a highly developed art which was practiced as successfully by the ancient Egyptians, the Saracens, and the Romans as by us. The present, however, is an age of science; we are not satisfied with rule-of-thumb methods, we want to know the *why* as well as the *how* of our practical operations. Only such knowledge of the reasons for methods empirically successful can enable us to drop out of our practice all superfluous steps and roundabout methods and to proceed straight to the mark in the most direct way. The industrial history of the last century is full of instances in 'which a knowledge of causes in relation to processes, i. e. a *scientific* knowledge, has shortened and improved practice in quite unexpected ways. So we may not doubt the ultimate value in practice of a science of breeding, if such a science can be created.

A beginning has been made during the last ten years, starting with the rediscovery of Mendel's law of heredity in 1900. This book will be concerned largely with the operations of that law.

CHAPTER I

THE DUALITY OF INHERITANCE

AT the outset we may with profit inquire what is meant by heredity. When a child resembles a parent or grandparent in some striking particular, we say it *inherits* such-and-such a characteristic from the parent or grandparent in question. *By heredity, then, we mean organic resemblance based on descent.*

Resemblances due to heredity may exist even between individuals not related as ancestor and descendant, as for example between uncle and nephew. Here the resemblance rests on the fact that uncle and nephew are both descended from a common ancestor, and they resemble each other simply because they have both inherited the same characteristic from that ancestor. This form of inheritance is sometimes spoken of as collateral in distinction from direct inheritance. In all cases alike community of descent is the basis of resemblances which can be ascribed to heredity, whether direct or collateral. Mother and child, no less than uncle and nephew, resemble each other because they have received a common inheritance from a common ancestor.

Three biological facts of fundamental importance to a right understanding of heredity were known imperfectly or not at all in the time of Darwin and Mendel. These are (1) the fertilization of the egg, (2) the maturation of the egg, which must precede its fertilization, and (3) the non-inheritance of '' acquired '' characters. These we may consider in order.

Every new organism is derived from a pre-existing organism, so far as our present experience goes. It may not have been so always. Indeed, on the evolution theory, we must suppose that living matter originally arose from lifeless, inorganic matter. But if it did, this may have occurred, and probably did occur, under physical conditions quite different from those now existing. At the present time the most exhaustive researches fail to reveal the occurrence of spontaneous generation, that is, the origin of living beings other than from pre-existing living beings.

In asexual methods of reproduction a new individual arises out of a detached portion of the parent individual. Such methods of origin are varied and interesting, but do not concern us at present. In all the higher animals and plants a new individual arises, by what we call a sexual process, from the union of two minute bodies called the reproductive cells. They are an egg-cell furnished by the mother and a sperm-cell furnished by the father.

There is a great difference in size between egg and sperm. The egg is many thousand times greater in bulk, as seen in Fig. 1, for example, yet the influence of each in heredity appears to be equal to that of the other. This fact shows unmistakably that the bulk of the reproductive cell is not significant in heredity. A large part of the relatively huge egg can have no part in heredity. It serves merely as food for the new organism, furnishing it with building material until such a time as it can begin to secure food for itself. The essential material, so far as heredity is concerned, is evidently found in egg and sperm alike. It is plainly small in amount and possibly consists merely in ferment-like bodies which ini-

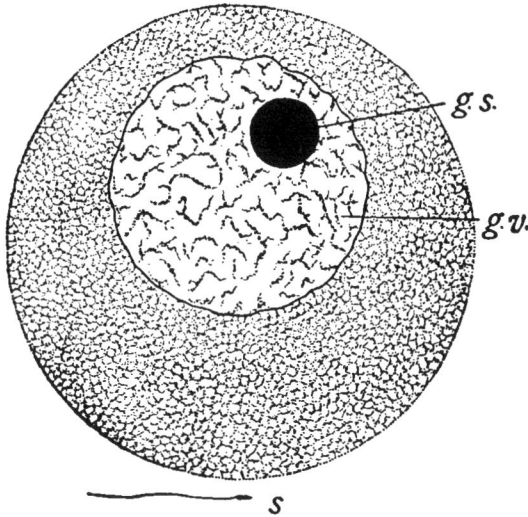

FIG. 1. — Egg and sperm (s) of the sea-urchin, *Toxopneustes*, both shown at the same enlargement. (After Wilson.)

tiate certain metabolic processes in a suitable medium represented by the bulk of the egg. The *amount* of a ferment used in starting a chemical change bears no relation, as is well known, to the amount of the chemical change which it can bring about in a suitable medium.

9

The equal share of egg and sperm in determining the character of offspring is well shown in the following experiment. An albino guinea-pig is one which lacks in large measure the ability to form black pigment. Apparently it does not possess some ingredient or agency necessary for the production of pigment. Now, if an albino male guinea-pig, such as is shown in Fig. 15, be mated with a black female guinea-pig of pure race, such as is shown in Fig. 14, young are produced all of which are black, like the mother, none being albinos, like the father. Fig. 16 shows black offspring produced in this way. Exactly the same result is obtained from the reverse cross, that is, from mating an albino mother with a black sire. It makes no difference, then, whether the black parent be mother or father, its blackness regularly dominates over the whiteness of the albino parent, so that only black offspring result. This fact, which has been repeatedly confirmed, shows that the black character is transmitted as readily through the agency of the minute sperm-cell as through the enormously greater egg-cell.

Let us now consider what happens when egg

and sperm unite, in what we call the fertilization of the egg. The egg is a rounded body incapable of motion, but the sperm is a minute thread-like body which moves like a tadpole by vibrations of its tail. In the case of most animals which live in the water, egg and sperm-cells are discharged into the water and there unite and develop into a new individual, but in the case of most land animals this union takes place within the body of the mother. We may consider an illustration of either sort.

The fertilization of the egg of a marine worm, Nereis, is shown in Fig. 2. The thread-like sperm penetrates into the egg. Its enlarged head-end forms there a small nuclear body, which increases in size until it equals that of the egg-nucleus, with which it then fuses. The egg next begins to divide up to form the different parts of a new worm-embryo. To each of these parts the nuclear material of egg and sperm is distributed equally. Since this development takes place wholly outside the body of either parent it is necessary that the egg contain enough food to last until the young worm can feed itself. This food material is

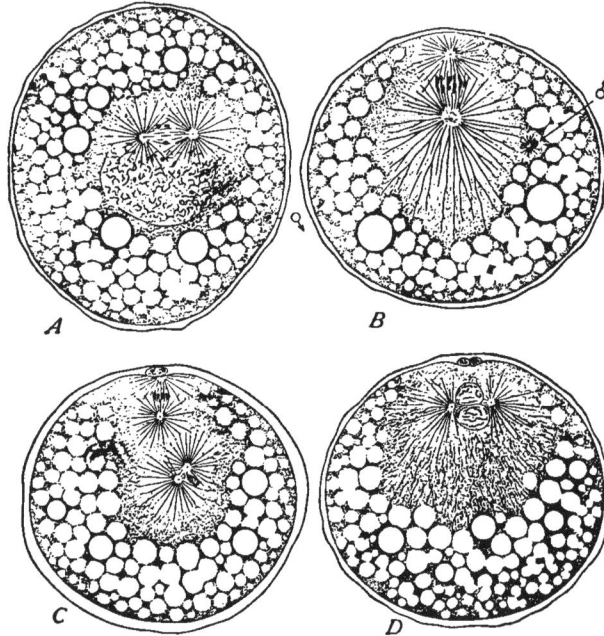

FIG. 2. — Fertilization of the egg of *Nereis*.

A. The sperm has entered the egg and is forming a minute nucleus at ♂. The egg-nucleus is breaking up preparatory to the first maturation division. B. The egg-nucleus is undergoing the first maturation division. Notice the conspicuous rod-like chromosomes separating into two groups. The sperm-nucleus (♂) is now larger and lies deeper in the egg. C. A small polar-cell has been formed above by the first maturation division of the egg. A second division is in progress at the same point. The sperm-nucleus is now deep in the egg and is preceded by a double radiation (amphiaster). D. Two polar-cells are fully formed. The matured egg-nucleus is now fusing with the sperm-nucleus. An amphiaster indicates that division of the egg will soon take place. (After Wilson.)

12

represented in part by the conspicuous oil-drops seen in the egg (the heavy circles in Fig. 2).

The egg of a mouse needs no such store of nourishment, since in common with the young of other mammals the mouse-embryo nourishes itself by osmosis from the body fluids of the mother. The mouse-egg is accordingly smaller. Stages in its fertilization are shown in Fig. 4. In *A* the sperm has already entered the egg. Remnants of its thread-like tail may still be seen there. Nearby is seen a nuclear body derived from the sperm-head. Opposite is seen the nuclear body furnished by the egg itself. The two nuclear bodies fuse and their united substance is then distributed to all parts of the embryo-mouse, just as happens in the development of the worm, Nereis.

There are reasons for thinking that the nuclear material is especially important in relation to heredity and that the equal share of the two parents in contributing it to the embryo is not without significance, for inheritance, as we have seen, is from both parents in equal measure. In cases where the inheritance from

13

each parent is different it can be shown that the offspring possess two inherited possibilities, though they may show but one. Thus in the case of a black guinea-pig, one of whose parents was white, the other black, it can be shown that the animal transmits both qualities (black and white) which it received from its respective parents, and transmits them in equal measure. For, if the cross-bred black animal be mated with a white one, half the offspring are black and half of them white. The cross-bred black animal inherited black from one parent, white from the other. It showed only the former, but on forming its reproductive cells it transmitted black to half of these, white to the other half. Hence the cross-bred black individual was a duality, containing two possibilities, black and white, but its reproductive cells were again single, containing either black or white, but not both.

Now it has been shown in recent years that the nuclear material in the reproductive cells behaves exactly as do black and white in the cross just described. This nuclear material becomes doubled in amount at fertilization,

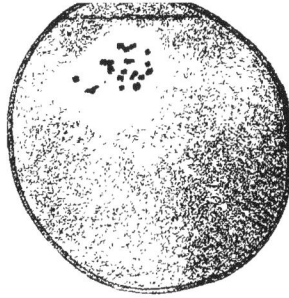

FIG. 3. — Egg of a mouse previous to maturation. (After Kirkham.)

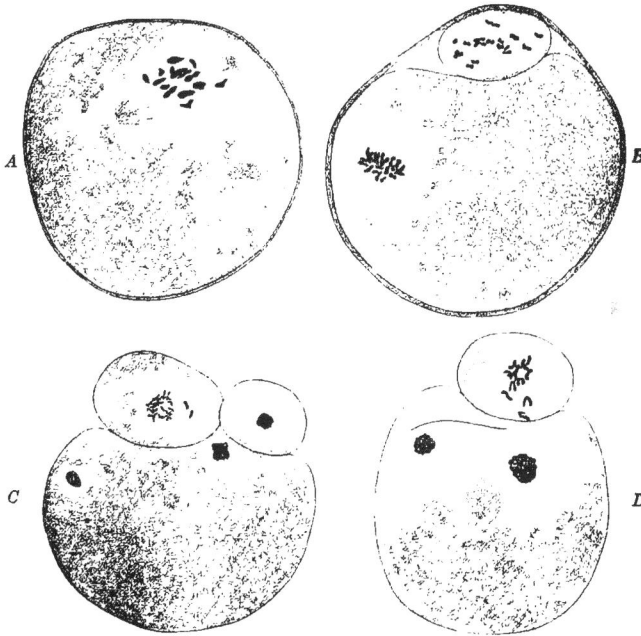

FIG. 4. — Maturation and fertilization of the egg of a mouse. A. The first maturation division in progress. B. The first polar-cell fully formed ; the second maturation division in progress. C. The second maturation division completed ; the second polar-cell is the smaller one ; near it, in the egg, is the egg-nucleus, and at the left is the sperm-nucleus. D. A view similar to the last, but showing only

equal contributions being made by egg and sperm. This double condition persists throughout the life of the new individual in all its parts and tissues. But if the individual forms eggs or sperm, these, before they can function in the production of a new individual, must undergo reduction to the single condition.

This reduction process is called maturation; it is well illustrated in the case of the mouse-egg, whose fertilization has already been described. The large nucleus of the egg-cell, as it leaves the ovary, is either broken up or about to break up preparatory to a cell-division. The most conspicuous of the nuclear constituents are some dense, heavily staining bodies called chromosomes, about twenty-four in number. In Fig. 3 each of these is split in two, preparatory to the first maturation division. The egg now divides twice, both times very unequally (Fig. 4), forming thus two smaller cells called polar cells, or polar bodies. They take no part in the formation of the embryo. The chromosomes left in the egg after these two divisions are only about half as numerous as before, or about twelve in number. These form the chro-

15

matin contribution of the egg to the production of a new individual. It is possible that other cell constituents undergo a similar reduction by half during maturation, but of this we have no present knowledge.

The known fact of chromosome reduction, of course, favors the current interpretation that the chromosomes are bearers of heredity, though it by no means proves the correctness of that interpretation. In the egg of Nereis, as well as in that of the mouse, two maturation divisions precede the fertilization of the egg. See Fig. 2. In *B* the first maturation division is in progress; in *C* the second is in progress; and in *D* both polar cells are fully formed, while egg and sperm nuclei are uniting. Similar processes occur in eggs generally, prior to their fertilization.

Like changes occur also in the development of the sperm-cells. In Fig. 5 the original or unreduced condition of the chromosomes in a cell of the male sexual gland is shown (at *A*) as one of four chromosomes to a cell. After a series of changes involving as in the maturation of the egg two cell-divisions, we find (at *H*) that the products of the original cell contain in each case two chromosomes, half the original number.

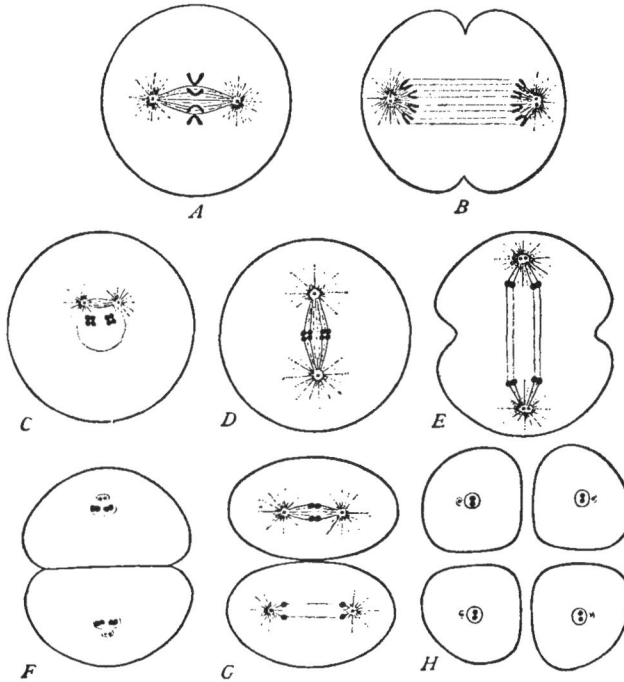

FIG. 5. — Diagrams showing the essential facts of chromosome reduction in the development of the sperm-cells. (After Wilson.)

These chromosomes make up the bulk of the head of the sperm which forms from each of these cells, its tail being derived from other portions of the cell.

It follows that not only eggs but also sperms, prior to their union in fertilization have passed into a reduced or single state as regards their chromatin constituents, whereas the fertilized egg, and the organism which develops from it, is in a double condition. It will be convenient to refer to the single condition as the N condition, the double as the 2 N condition.

From a wholly different source we have evidence strongly confirmatory of the conclusion that the fertilized egg contains a double dose of the essential nuclear material. By artificial means it has been found possible to cause the development of an unfertilized egg. The means employed may be of several different sorts, such as stimulation with acids, alkalies, or solutions of altered density. In such ways the development has been brought about of the eggs of sea-urchins, star-fishes, worms, and mollusks, which normally require fertilization to make them develop.

The sea-urchin egg has been made to develop more successfully than any other. This has occurred even after the egg had undergone maturation, being reduced to the N condition. From the development of such reduced but unfertilized eggs fully normal sea-urchins have been obtained which even contain developed sexual glands. On the other hand it has been found possible to break the egg into fragments by shaking it, or cutting it into bits with fine knives or scissors. It has also been found possible to bring about the development of an egg fragment so obtained, — a fragment which contained no egg nucleus. This result has been attained by allowing a sperm to enter it and form there a nuclear body. No adult organism has yet been reared from such a fertilized egg-fragment, but so far as the development has been followed it progressed normally.

There can accordingly be no doubt that the nuclear material of a sperm-cell has all the capabilities of that of an egg-cell and can indeed replace it in development. Accordingly, when, as in normal fertilization, both an egg nucleus and a sperm nucleus are present in the cell, a double dose of the necessary nuclear

19

material is supplied. The second or extra dose is, however, not superfluous. It probably adds to the vigor of the organism produced, and in some cases at least, materially affects its form. For many animals and plants exist in two different conditions, in one of which the nuclear components are simple, N, while in the other they are double, 2 N. Thus in bees, rotifers, and small crustacea the egg may under certain conditions develop without being fertilized. If the egg develops before maturation is complete, that is in the 2 N condition, the animal produced is a female, like the mother which produced the egg. But if the egg undergoes reduction to the N condition before beginning its development, then it produces a male individual, an organism, so far as reproduction is concerned, of lower metabolic activity.

In many plants, too, individuals of N and of 2 N constitution occur, which differ markedly in appearance. Thus the ordinary fern-plant is a 2 N individual, but it never produces 2 N offspring. Fig. 6 shows an ordinary fern-plant, which produces spores on the under

Fig. 6. — An ordinary fern, which reproduces by asexual spores. The fern is shown reduced in size at 382; a portion of a frond seen from below and slightly enlarged, at 383; a cross-section of the same more highly magnified, at 384. Notice in 384 the sporangia, and in 385 one of these discharging spores. (After Wossidlo, from Coulter Barnes and Cowle's Textbook of Botany.)

surface of its fronds. Each of those spores is a reproductive cell which, like the mature eggs and sperm of animals, is in a reduced nuclear condition (N). These spores germinate, however, without uniting in pairs and form a plant different from the parent, just as the mature egg of a bee, if unfertilized, develops into an individual different from the parent, in that case a male. The plant which develops from the spore of a fern is small and inconspicuous and is known as a prothallus. See Fig. 7. It produces sexual cells (eggs and sperm) which, uniting in pairs, form fern-plants, 2 N individuals. Thus there is a constant alternation of generations, fern-plants (2 N), which produce prothalli (N), and then these produce again fern-plants (2 N).

The fact is worthy of note that in an animal or plant which is in the single or N condition, there occurs no chromatin reduction at the formation of reproductive cells. Its cells are already in the single condition, and they probably cannot be further reduced without destroying the organism. The 2 N fern-plant forms reproductive cells, its spores, which are in the reduced condition, N, and these germinate into the prothallus, which accordingly is

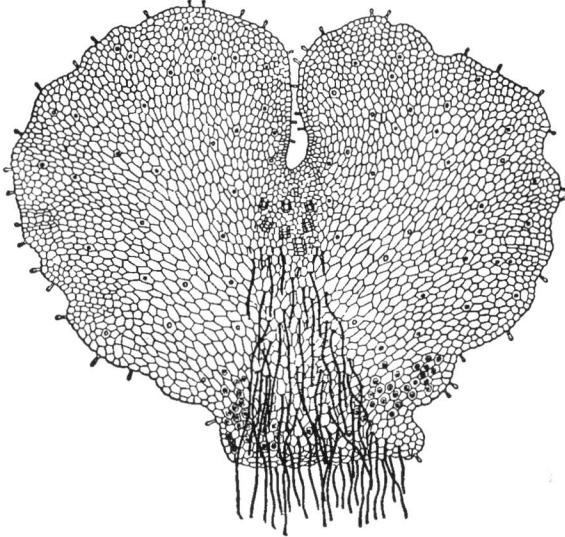

FIG. 7. — The prothallus of a fern, which reproduces by sexual cells, eggs and sperm. The eggs are borne in the sac-like "archegonia," just below the notch in the figure. They, like the sperm-forming "antheridia," lie on the under surface of the flattened prothallus which is here viewed from below. Notice the root-hairs or rhizoids by which the plant feeds. Highly magnified. (After Coulter, Barnes, and Cowles.)

N throughout. But when the prothallus forms reproductive cells, no reduction occurs. Its egg-cells and its sperm-cells in common with

all other cells of the prothallus are already
in the reduced condition without any maturation divisions. The result of their union in
pairs, at fertilization, is the formation of 2 N
combinations that germinate into fern-plants.

Similarly in the case of a male animal which

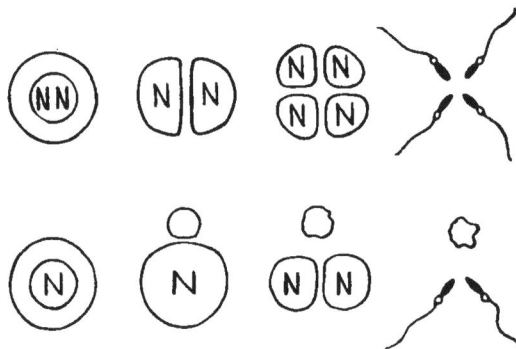

FIG. 8. — Diagram showing the chromosome number in the
spermatogenesis of ordinary animals (upper line) and of the
wasp (lower line).

has developed from a reduced but unfertilized
egg, no reduction occurs at the formation of
its sperm-cells. In an ordinary male animal,
one which is in the double or 2 N state, the
development of the sperms is attended by reduction to the N condition. In this process
there occur two cell-divisions producing from
each initial cell four sperms. See Fig. 5, and

Fig. 8, upper line. But in the male wasp, whose cells are in the N condition at the beginning, one of these divisions is so far suppressed that the resulting cell products are of very unequal size, and the smaller one contains no nuclear material. The other then gives rise to two sperm-cells, each possessing the original N nuclear condition, while the small non-nucleated cell degenerates. See Fig. 8, lower line.

In conclusion, I wish to introduce two technical terms, which it will be convenient for us to use in subsequent discussions. These are *gamete* and *zygote*. A reproductive cell (either egg or sperm) which is in the reduced condition (N) ready for union in fertilization is called a gamete. The result of fertilization is a zygote, a joining together of two cells each in the N condition. The result is a new organism, at first a single cell, in the 2 N condition.

BIBLIOGRAPHY

CASTLE, W. E.
1903. "The Heredity of Sex." *Bull. Mus. Comp. Zoölogy*, 40, pp. 189–218.

DELAGE, Y.
1898. "Embryons sans noyau maternel." *Compte rendu, Académie des sciences, Paris*, 127, pp. 528–531.
1909. "Le sexe chez les Oursins issus de parthénogenèse experimentale." *Compte rendus, Académie des sciences, Paris*, 148, pp. 453–455.

KIRKHAM, W. B.
1907. Maturation of the Egg of the White Mouse." *Trans. Conn. Acad. of Arts and Sciences*, 13, pp. 65–87.

LOEB, J.
1899. "On the Nature of the Process of Fertilization and the Artificial Production of Normal Larvae (Plutei) from the Unfertilized Eggs of the Sea-urchin." *Amer. Journ. of Physiol.*, 3, pp. 135–138.

LOTSY, J. P.
1905. "Die X-Generation und die 2 X-Generation." *Biologisches Centralblatt*, 25, pp. 97–117.

MEVES, F., und DUESBERG, J.
1908. "Die Spermatozytenteilungen bei der Hornisse (Vespa crabo L.)." *Arch. f. mik. Anat. u. Entwick.*, 71, pp. 571–587.

WILSON, E. B.
1896. "The Cell in Development and Inheritance," 370 pp., illustrated. The Macmillan Co., New York.

CHAPTER II

IN the last chapter we discussed two biological principles which, if clearly grasped,
greatly simplify an understanding of the
process of heredity. These are as follows:

(1) A sexually produced individual arises
from the union of two reproductive cells (or
gametes), each of which contains, so far as
heredity is concerned, a full material equipment for the production of a new individual.
Accordingly, the newly produced individual is
two-fold or duplex as concerns the material
basis of heredity.

(2) If the new individual becomes adult and
forms gametes, the production of these will be
attended by a reduction to the simplex or
single condition as regards the material basis
of heredity.

27

To these two principles we may now add a third, viz.: — (3) The individual consists of two distinct parts: first, its body destined to die and disintegrate after a certain length of time; and, secondly, the germ-cells contained within that body, capable of indefinite existence in a suitable medium.

The fertilized egg or zygote begins its independent existence by dividing into a number of cells. These become specialized to form the various parts and tissues of the body, muscle, bone, nerve, etc., and by becoming thus specialized they lose the power to produce anything but their own particular kind of specialized tissue; they cannot reproduce the whole. This function is retained only by certain undifferentiated cells found in the reproductive glands and known as germ-cells. They are direct lineal descendants of the fertilized egg itself. If they are destroyed the individual loses the power of reproduction altogether.

External influences which act upon the body may of course modify it profoundly, but such modifications are not transmitted through the gametes, because the gametes are not derived from body-cells, but from germ-cells. This relationship first pointed out by Weismann may be expressed in a diagram, as in Fig. 9. Only such environmental influences as directly alter the character of the germ-cells will in any way influence the character of subsequent generations of individuals derived from those

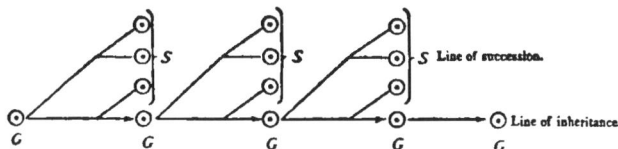

FIG. 9. — Diagram showing the relation of the body (S) to the germ-cells (G) in heredity. (After Wilson.)

germ-cells. Body (or somatic) influences are not inherited. This knowledge we owe largely to Weismann, who showed experimentally that mutilations are not inherited. The tails of mice were cut off for twenty generations in succession, but without effect upon the character of the race. Weismann also pointed out the total lack of evidence for the then current belief that characters acquired by the body are inherited. The correctness of his view that body and germ-cells are physiologically distinct

29

is indicated by the results obtained when germ-cells are transplanted from one individual to another.

Heape showed some twenty years ago that if the fertilized egg of a rabbit of one variety (for example an angora, i. e. a long-haired, white animal) be removed from the oviduct of its mother previous to its attachment to the uterine wall, and be then transferred to the oviduct of a rabbit of a different variety (for example a Belgian hare, which is short-haired and gray), the egg will develop normally in the strange body and will produce an individual with all the characteristics of the real (angora) mother unmodified by those of the foster mother (the Belgian hare). Young thus obtained by Heape were both long-haired and albinos, like the angora mother. To this experiment the objection might be offered that the transplanted egg was already full-grown and fertilized when the transfer was made, and that therefore no modification need be expected, but if the egg were transferred at an earlier stage the result might have been different. In answer to such a possible objection the follow-

FIG. 10.

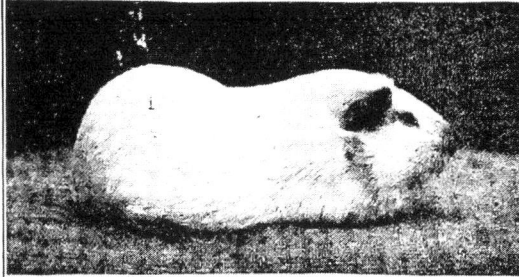

FIG. 11.

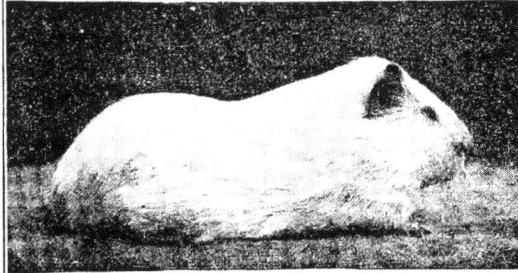

FIG. 12.

FIG. 10. — A young, black guinea-pig, about three weeks old. Ovaries taken from an animal like this were transplanted into the albino shown below.

FIG. 11. — An albino female guinea-pig. Its ovaries were removed, and in their place were introduced ovaries from a young, black guinea-pig, like that one shown in Fig. 10.

FIG. 12. — An albino male guinea-pig, with which was mated the albino shown in Fig. 11.

ing experiment performed by Dr. John C. Phillips and myself may be cited.

A female albino guinea-pig (Fig. 11) just attaining sexual maturity was by an operation deprived of its ovaries, and instead of the removed ovaries there were introduced into her body the ovaries of a young black female guinea-pig (Fig. 10), not yet sexually mature, aged about three weeks. The grafted animal was now mated with a male albino guinea-pig (Fig. 12). From numerous experiments with albino guinea-pigs it may be stated emphatically that normal albinos mated together, without exception, produce only albino young, and the presumption is strong, therefore, that had this female not been operated upon she would have done the same. She produced, however, by the albino male three litters of young, which together consisted of six individuals, all black. (See Fig. 13.) The first litter of young was produced about six months after the operation, the last one about a year. The transplanted ovarian tissue must have remained in its new environment therefore from four to ten months before the eggs attained full growth and were discharged, ample time, it would seem, for the influence of a foreign body upon the inheritance to show itself were such influence possible.

In the light of the three principles now stated, viz. (1) the duplex condition of the zygote, (2) the simplex condition of the gametes, and (3) the distinctness of body and germ-cells, we may proceed to discuss the greatest single discovery ever made in the field of heredity, — Mendel's law.

BIBLIOGRAPHY

CASTLE, W. E., and PHILLIPS, JOHN C.
  1911. "On Germinal Transplantation in Vertebrates." *Carnegie Institution of Washington, Publication No. 144*, 26 pp., 2 pl.

HEAPE, W.
  1890. "Preliminary Note on the Transplantation and Growth of Mammalian Ova within a Uterine Foster-mother." *Proc. Roy. Soc.*, 48, pp. 457–458.
  1897. "Further Note," etc. *Id.* 62, pp. 178–183.

WEISMANN, A.
  1893. "The Germ-Plasm." Translation by Parker and Römfeldt. Chas. Scribner's Sons, New York.

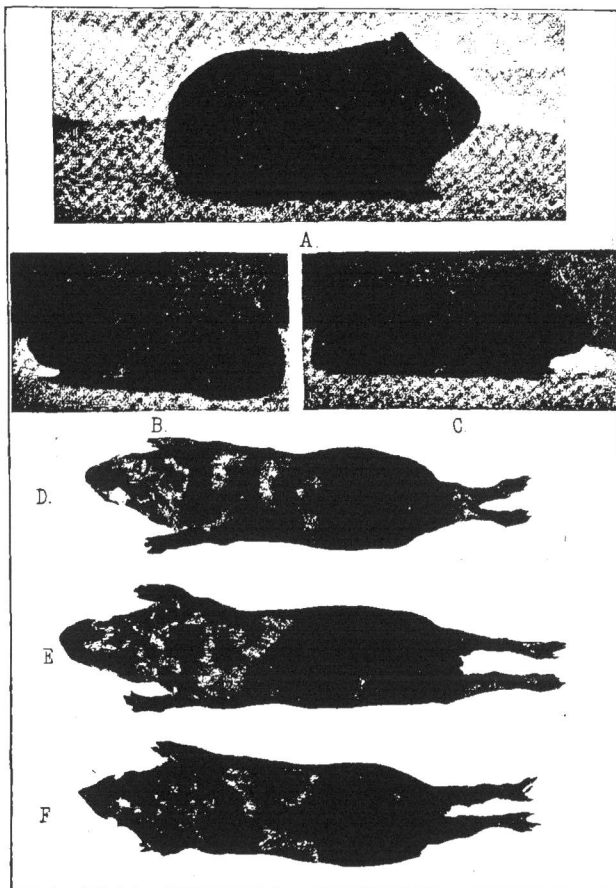

FIG. 13. — Pictures of three living guinea-pigs (*A, B, C*), and of the preserved skins of three others (*D, E, F*); all of which were produced by the pair of albinos shown in Figs. 11 and 12.

CHAPTER III

MENDEL'S LAW OF HEREDITY

GREGOR JOHANN MENDEL was a teacher of the physical and natural sciences in a monastic school at Brünn, Austria, in the second half of the last century. He was, therefore, a contemporary of Darwin, but unknown to him as to nearly all the great naturalists of the period. Although not famous in his lifetime, it is clear to us that he possessed an analytical mind of the first order, which enabled him to plan and carry through successfully the most original and instructive series of studies in heredity ever executed. The material which he used was simple. It consisted of garden-peas, which he raised in the garden of the monastery. The conclusions which he reached were likewise simple. He summed them up, the results of eight years of arduous work, in a brief paper published in the proceedings of the local

scientific society. There they remained un-
heeded for thirty-four years, until their author
had long been dead. Meantime biological science had made steady progress. It reached
the position Mendel had attained in advance
of his time, and Mendel's law was rediscovered simultaneously in 1900 by De Vries in
Holland, by Correns in Germany, and by
Tschermark in Austria. It gratifies our sense
of poetic justice that to-day the rediscovered
law bears the name, not of any one or of all of
its brilliant rediscoverers, but of the all-but-
forgotten Mendel.

The essential features of this law can best
be explained in connection with some illustrations, which I choose for convenience from my
own experiments. If a black guinea-pig of
pure race (Fig. 14) be mated with a white one
(Fig. 15), the offspring will, as explained on
page 10, all be black; none will be white.
To use Mendel's terminology, the black character dominates in the cross, while white
recedes from view. The black character is,
therefore, called the *dominant* character;
white, the *recessive* character.

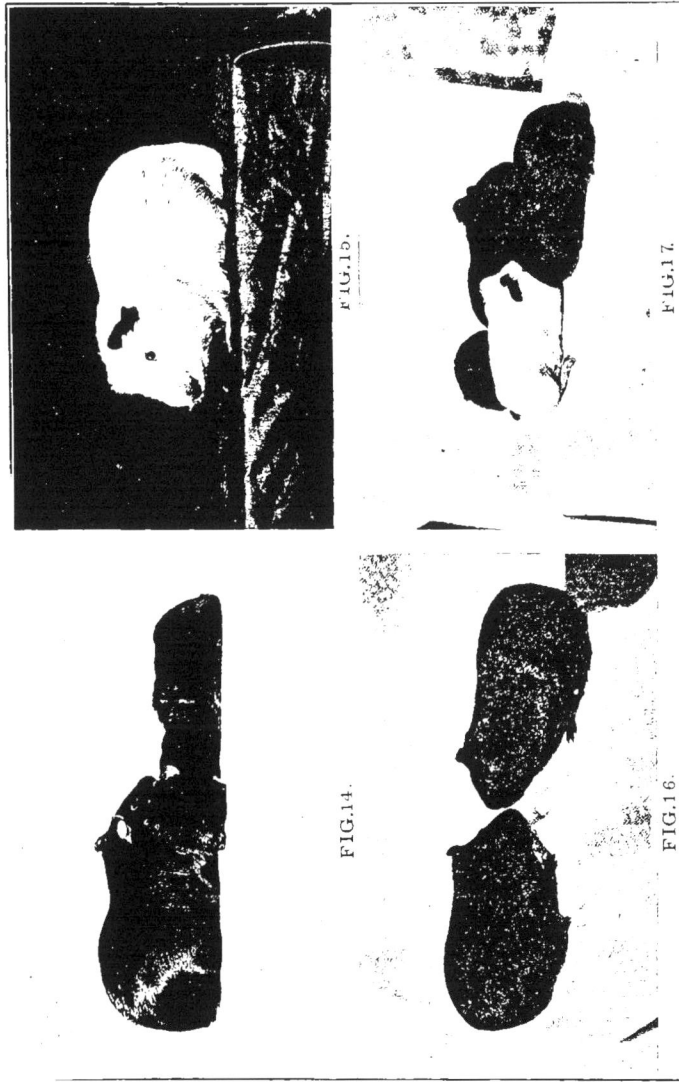

FIG. 14. — A black, female guinea-pig, and her young.

FIG. 15. — An albino male guinea-pig, father of black young like those shown in Fig. 14.

FIG. 16. — Two of the grown-up young of a black and of an albino guinea-pig. Compare Figs. 14 and 15.

FIG. 17. — A group of four young, produced by the animals shown in Fig. 16.

But, if now two of the cross-bred black individuals (Fig. 16) be mated with each other, the recessive white character reappears on the average in one in four of the offspring (Fig.

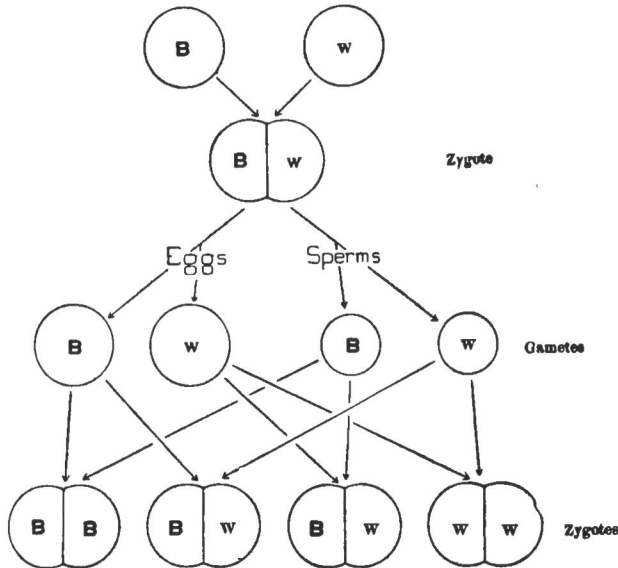

FIG. 18. — Diagram to explain the result shown in Fig. 17.

17). Its reappearance in that particular proportion of the offspring may be explained as follows (see Fig. 18): The gametes which united in the original cross were, one black, the other white in character. Both characters were then asociated together in the offspring; but black from its nature dominated, because white in this case is due merely to the lack of some constituent supplied by the black gamete. But when the cross-bred black individuals on becoming adult form gametes, the black and the white characters separate from each other and pass into different cells, since, as we have seen, gametes are simplex. Accordingly, the eggs formed by a female cross-bred black are half of them black, half of them white in character, and the same is true of the sperms formed by a male cross-bred black. The combinations of egg and sperm which would naturally be produced in fertilization are accordingly 1 B B : 2 B W : 1 W W, or three combinations containing black to one containing only white, which is the ratio of black to white offspring observed in the experiment.

Now the white individual may be expected to transmit only the white character, never the black, because it does not contain that character. Experiment shows this to be true. White guinea-pigs mated with each other produce only white offspring. But the black in-

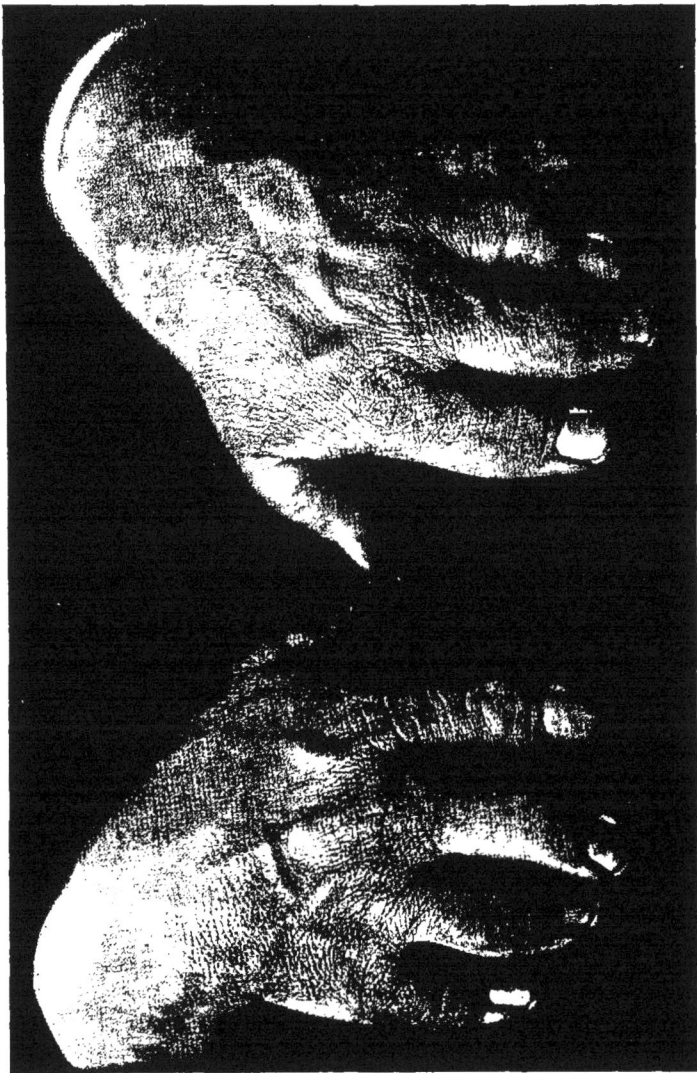

FIG. 19. — A shortened condition of the skeleton, particularly of the fingers, as here shown, is a dominant character in heredity. (After Farabee.)

dividuals of this generation are of two sorts,
— B B and B W in character. The B B individual is *pure*, so far as its breeding capacity
is concerned. It can form only black (B)
gametes. But the B W individuals may be
expected to breed exactly like the cross-bred
blacks of the previous generation, forming
gametes, half of which will carry B, half W.
Experiment justifies both these expectations.
The test may readily be made by mating the
black animals one by one with white ones.
The pure (or B B) black individual will produce only black offspring, whereas those not
pure, but B W in character, will produce offspring half of which on the average will be
black, the other half white. These two kinds
of dominant individuals obtained in the second
generation from a cross we may for convenience call homozygous and heterozygous, following the convenient terminology of Bateson.
A homozygous individual is one in which *like*
characters are joined together, as B with B;
a heterozygous individual is one in which *unlike*
characters are joined together, as B with W.
It goes without saying that recessive individ-

uals are always homozygous, as W W for example. For they do not contain the dominant character, otherwise they would show it.

It will be observed that in the cross of black with white guinea-pigs black and white behave as units distinct and indestructible, which may meet in fertilization but separate again at the formation of gametes. Mendel's law as illustrated in this cross includes three principles: (1) The existence of *unit-characters,* (2) *dominance,* in cases where the parents differ in a unit-character, and (3) *segregation* of the units contributed by the respective parents, this segregation being found among the gametes formed by the offspring.

The principles of dominance and segregation apply to the inheritance of many characteristics in animals and plants. Thus in guinea-pigs a rough or rosetted coat (Figs. 23 and 24) is dominant over the ordinary smooth coat. If a pure rough individual is crossed with a smooth one, all the offspring are rough; but in the next generation smooth coat reappears in one fourth of the offspring, as a rule. Again, in guinea-pigs and rabbits a long or angora condition of the

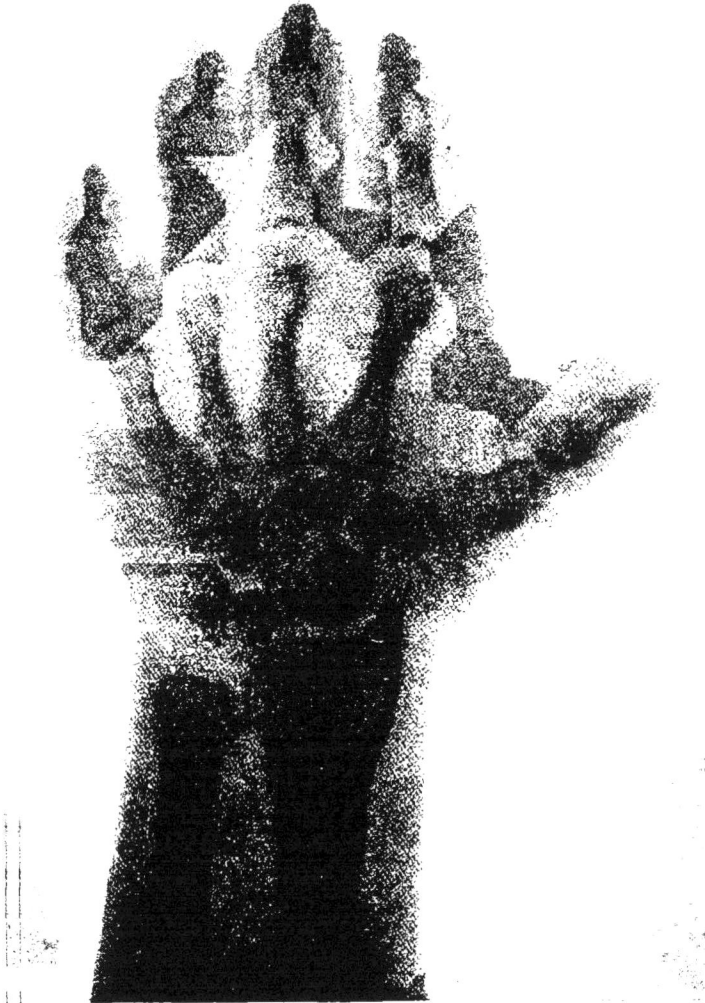

FIG. 20. — Radiograph of a hand similar to those shown in Fig. 19. Notice the short, two-jointed fingers. (After Farabee.)

fur is recessive in crosses with normal short hair. All the immediate offspring of such a cross are short-haired, but in the next generation long hair reappears in approximately one fourth of the offspring.

In cattle, the polled or hornless condition is dominant over the normal horned condition; in man, two-jointed fingers and toes (Figs. 19 and 20) are dominant over normal three-jointed ones. This is clear from an interesting pedigree given by Farabee of the inheritance of the abnormality in a Pennsylvania family (see Fig. 21). In no case was an abnormal member of the family known to have married any but an unrelated normal individual. It will be seen that approximately half the offspring throughout the four generations of offspring shown in the table were of the abnormal sort, — short-bodied and with short fingers and toes.

In each of the cases thus far considered a single unit-character is concerned. Crosses in such cases involve no necessary change in the race, but only the continuance within it of two sharply alternative conditions. But the result is quite different when parents are crossed

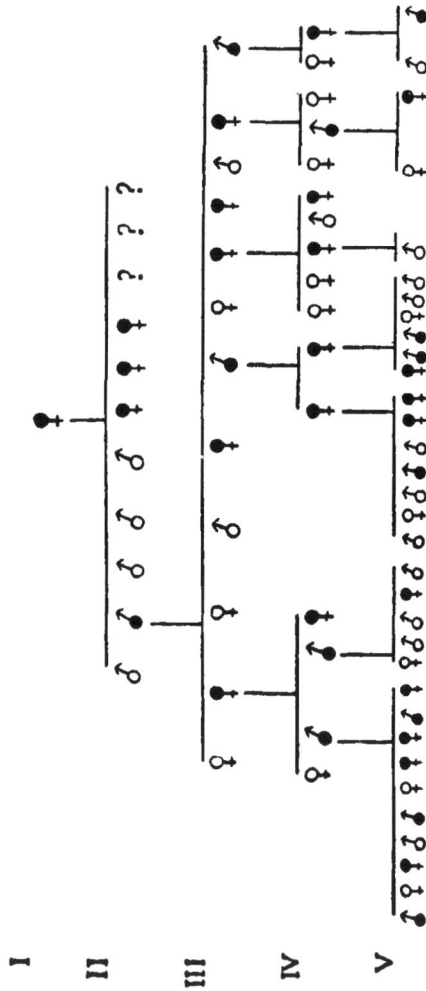

FIG. 21. — Diagram showing the descent, through five generations, of the condition shown in Figs. 19 and 20. Black symbols indicate affected individuals.

FIG. 22.

FIG. 23.

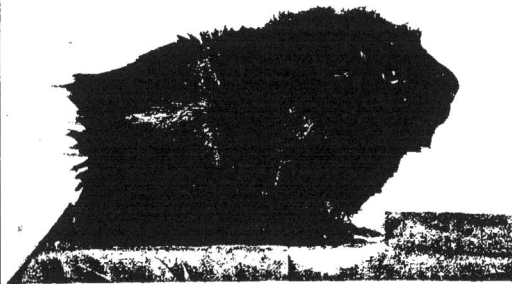

FIG. 24.

FIG. 22. — A smooth, dark guinea-pig.

FIG. 23. — A rough, white guinea-pig.

FIG. 24. — A dark, rough guinea-pig. The new combination of characters obtained when animals are mated like those shown in Figs. 22 and 23.

which differ simultaneously in two or more independent unit-characters. Crossing them becomes an active agency for the production of new varieties.

In discussing the crosses now to be described it will be convenient to refer to the various generations in more precise terms, as Bateson has done. The generation of the animals originally crossed will be called the parental generation (P); the subsequent generations will be called filial generations, viz. the first filial generation (F_1), second filial (F_2), and so on.

When guinea-pigs are crossed of pure races which differ simultaneously in two unit-characters, the F_1 offspring are all alike, but the F_2 offspring are of four sorts. Thus, when a smooth dark animal (Fig. 22) is crossed with a rough white one (Fig. 23) the F_1 offspring are all rough and dark (Fig. 24), manifesting the two dominant unit-characters, — dark coat derived from one parent, rough coat derived from the other. But the F_2 offspring are of four sorts, viz. (1) smooth and dark, like one grandparent, (2) rough and white, like the other grandparent, (3) rough and dark, like

41

the F_1 generation, and (4) smooth and white, a new variety (Fig. 25). It will be seen that the pigmentation of the coat has no relation to its smoothness. The dark animals are either rough or smooth, and so are the white ones. Pigmentation of the coat is evidently a unit-character independent of hair-direction, and as new combinations of these two units the cross has produced two new varieties, — the rough dark and the smooth white.

Again, hair-length is a unit-character independent of hair-color. For if a short-haired dark animal (either self or spotted, Fig. 26) be crossed with a long-haired albino (Fig. 27), the F_1 offspring are all short-haired and dark (Fig. 28); but the F_2 offspring are of four sorts, viz. (1) dark and short-haired, like one grandparent, (2) white and long-haired, like the other, (3) dark and long-haired, a new combination (Fig. 29), and (4) white and short-haired, a second new combination (compare Fig. 25).

Now the four sorts of individuals obtained from such a cross as this will not be equally numerous. As we noticed in connection with

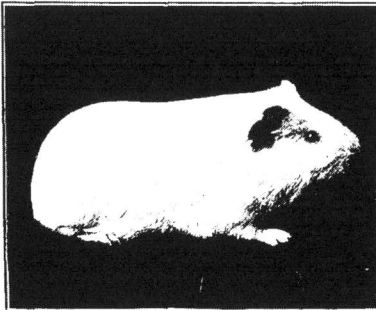
FIG. 25.

FIG. 26.

FIG. 27.

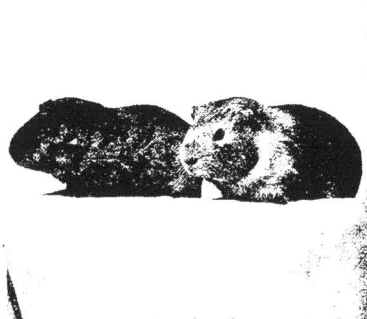
FIG. 28.

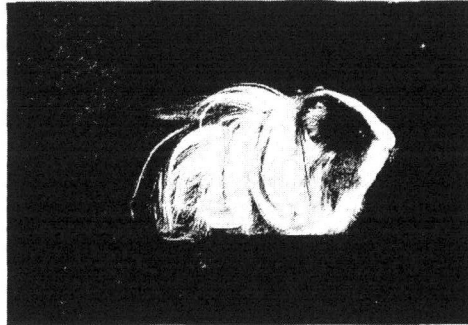
FIG. 29.

Fig. 25. — A smooth, white guinea-pig. A second new combination of characters, but obtained first among the *grandchildren* of such animals as are shown in Figs. 22 and 23.

Fig. 26. — A short-haired, pigmented guinea-pig. ("Dutch-marked" with white.)

Fig. 27. — A long-haired, albino guinea-pig

Fig. 28. — Offspring produced by animals of the sorts shown in Figs. 26 and 27. One shows the "Dutch-marked" pattern as a belt of pale yellow; the other does not. Both are short-haired and pigmented (not albinos).

Fig. 29. — A long-haired, pigmented guinea-pig, "Dutch-marked" with white. Its parents were like the animals shown in Fig. 28; its grandparents like those shown in Figs. 26 and 27.

the black-white cross, dominant individuals are to the corresponding recessives as three to one. Therefore, we shall expect the short-haired in-

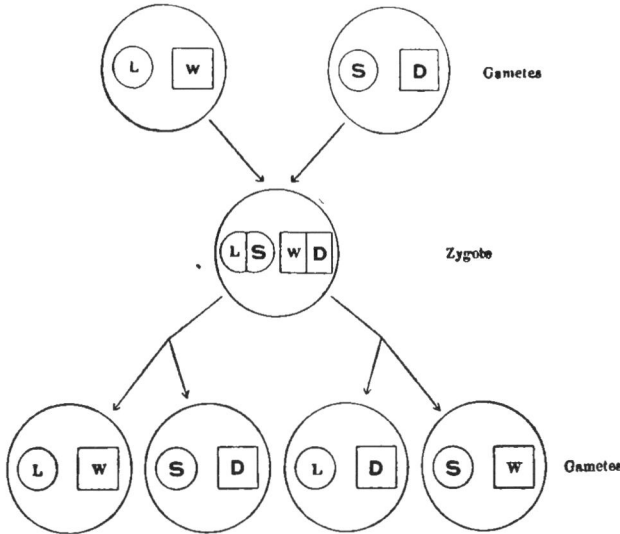

FIG. 30. — Diagram to explain the result of a cross between the sorts of guinea-pigs shown in Figs. 26 and 27. L stands for long hair, S for short hair, D for dark hair, and W for white hair. Dominant characters are indicated by heavy type.

dividuals in F_2 to be three times as numerous as the long-haired ones, and dark ones to be three times as numerous as white ones. Further, individuals which are *both* short-haired and dark should be 3×3 or 9 times as numerous as those which are not. The expected proportions of the four classes of F_2 offspring are accordingly $9:3:3:1$, a proportion which is closely approximated in actual experience. The Mendelian theory of independent unit-characters accounts for this result fully. No other hypothesis has as yet been suggested which can account for it.

Suppose that each unit has a different material basis in the gamete. Let us represent the material basis of hair-length by a circle, that of hair-color by a square, then combinations and recombinations arise as shown in Fig. 30. The composition of the gametes furnished by the parents is shown in the first line of the figure; that of an F_1 individual (or zygote), in the second line; that of the gametes formed by the F_1 individual in the third line. L meets S and W meets D in fertilization to form an F_1 individual double and also heterozygous as regards hair length and hair color, but these units segregate again as the gametes of the F_1 individuals are formed, and it is a matter of chance whether or not they are associated

44

as originally, L with W and S with D, or in a new relationship, L with D and S with W. Hence we expect the F_1 individuals to form four kinds of gametes all equally numerous, — L W, S D, L D, and S W. By chance unions of these in pairs nine kinds of combinations become possible, and their chance frequencies will be as shown in Fig. 31. Four of these combinations, including nine individuals, will show the two dominant characters, short and dark; two classes, including three individuals, will show one dominant and one recessive character, viz. dark and long; two more classes, including three individuals, will show the other dominant and the other recessive character, viz. short and white; and lastly, one class, including a single individual, will show the two recessive characters, long and white. The four *apparent* classes, or, as Johannsen calls them, *phenotypes*, will accordingly be as $9:3:3:1$. This is called the normal Mendelian ratio for a dihybrid cross, — that is, a cross involving two unit-character differences.

One individual in each of these four classes will, if mated with an individual like itself,

45

breed true, for it is homozygous, containing only like units. The double recessive class, long white, of course contains *only* homozygous individuals, but in each class which shows a dominant unit, heterozygous individuals outnumber homozygous ones, as $2:1$ or $8:1$. Now the breeder who by means of crosses has pro-

Short Dark.	Long Dark.	Short White.	Long White.
1 S D. S D	1 L D. L D	1 S W. S W	1 L W. L W
2 S D. L D	2 L D. L W	2 S W. L W	
2 S D. S W			
4 S D. L W			
9	3	3	1

Fig. 31. — Diagram showing the kinds and relative frequencies of the young to be expected in F_2 from the crossing of animals shown in Figs. 26 and 27.

duced a new type of animal wishes, of course, to " fix " it, — that is, to obtain it in a condition which will breed true. He must, therefore, obtain homozygous individuals. If he is dealing with a combination which contains only recessive characters, this will be easy enough, for such combinations are invariably homozygous. His task will become increasingly difficult the more dominant characters there are included in the combination which he desires to fix.

46

Fig. 32. — A long-haired, rough albino guinea-pig ;
male, 2002.

The most direct method for him to follow is to test by suitable matings the unit-character constitution of each individual which shows the desired combination of characters, and to reject all which are not homozygous. In this way a pure race may be built up from individuals proved to be pure. Such a method, however, though sure, is slow in cases where the desired combination includes two or more dominant unit-characters, for it involves the application of a breeding test to many dominant individuals, most of which must then be rejected. It is therefore often better in practice to breed from all individuals which show the desired combination, and eliminate from their offspring merely such individuals as do not show that combination. The race will thus be only gradually purified, but a large stock of it can be built up much more quickly.

We may next discuss a cross in which three unit-character differences exist between the parents, instead of two. If guinea-pigs are crossed which differ simultaneously in three unit-characters, color, length, and direction of the hair, a still larger number of phenotypes is obtained in F_2, namely, eight. A cross between a short-haired, dark, smooth guinea-pig (compare Fig. 22) and one which was long-haired, white, and rough (Fig. 32) produced offspring in F_1 which were short-haired, dark, and rough (compare Fig. 24), these being the three dominant characters, two derived from one parent, one from the other. The F_2 offspring were of eight distinct types, two like the respective grandparents, one like the F_1 individuals (parents), and the other five new, shown in Fig. 33. They are short white rough, short white smooth, long white smooth, long dark rough, and long dark smooth. The largest of the eight apparent classes (phenotypes) was the one which manifested the three dominant characters, short, dark, and rough, which had been the exclusive F_1 type; the smallest class was the one which manifested the three recessive characters, long, white, and smooth. Theoretically these two classes should be to each other as $27:1$. Of the twenty-seven triple-dominants, twenty-six should be heterozygous.

A comparison of this case with the one just previously described shows what an increas-

FIG. 33. — Five new combinations of unit-characters obtained in generation F_2, by crossing the animal shown in Fig. 32 with animals like that shown in Fig. 22.

ingly difficult thing it is to fix types obtained by crossing, if the number of dominant characters

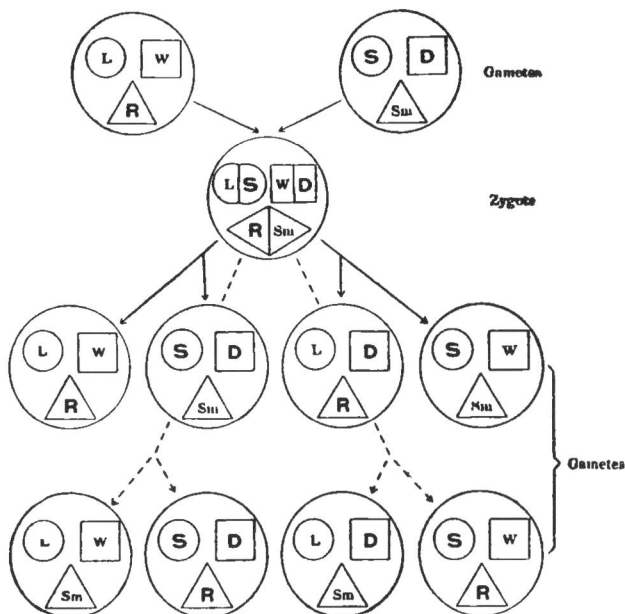

FIG. 34. — Diagram to show the gametic combinations and segregations involved in a cross between guinea-pigs differing in three unit-characters. L stands for long hair, S for short hair, W for white hair, and D for dark hair; R for rough, and Sm for smooth coat. Compare Figs. 22 and 32.

in the selected type increases. On the theory of unit-characters the gametic combinations and segregations involved in this cross are as shown in Fig. 34. The nature of the gametes formed by the parents crossed is shown in the first row; the composition of the F_1 individuals, immediately below. In the two lower rows are shown four different sorts of gametic splittings which may occur in the F_1 individuals, producing thus eight different kinds of gametes. If, in reality, the F_1 individuals form eight kinds of gametes, all equally numerous, and chance unions in pairs occur among them, there should be produced eight corresponding sorts of individuals numerically as $27:9:9:9:3:3:3:1$. In a total of 64 individuals there should be on the average one pure individual in each of the eight different classes. The class numerically 27 in 64 manifests three dominant characters; those which are numerically 9 in 64 manifest two dominant characters; those which are numerically 3 in 64 manifest one dominant character. Among each of these there will be on the average one pure individual, but the class which contains 1 individual in 64 is a pure recessive, for it contains no dominant character. This combination, then, requires no fixation. It will breed true from the start.

BIBLIOGRAPHY

CASTLE, W. E.

1903. "Mendel's Law of Heredity." *Proc. Am. Acad. Arts and Sci.*, 38, pp. 535–548; also, *Science, N. S.*, 18, pp. 396–406.

1905. "Heredity of Coat-characters in Guinea-pigs and Rabbits." *Carnegie Institution of Washington, Publication No. 23*, 78 pp., 6 pl.

FARABEE, W. C.

1905. "Inheritance of Digital Malformations in Man." *Papers of the Peabody Museum, Harvard University*, vol. 3, No. 3, pp. 69–77, 5 pl.

CHAPTER IV

W E have noticed that when a black
guinea-pig of pure race is mated
with a white one, only black offspring are produced; and that when rough-
coated guinea-pigs are mated with smooth-
coated ones, only rough-coated young are produced; and that when short-haired guinea-
pigs are mated with long-haired ones, only
short-haired young are produced. The character which in each case is seen in the young
we call dominant, that which is unseen we
call recessive. Thus black is dominant over
white, rough coat over smooth coat, and short
coat over long coat.

A question which has given much concern to
students of heredity is this, — upon what does

dominance depend? Why should black dominate white rather than the reverse?

In poultry, indeed, the relations are often reversed, white dominating black. Why is this? Several attempted explanations have been made, but none of them is thoroughly satisfactory. The one which has found widest acceptance is this: In the dominant individual something is present which is wanting in the recessive. Thus, in the black guinea-pig there is present some ferment-like body or some ingredient of black which is wanting in the albino. Accordingly, the black guinea-pig forms pigment, a thing which the albino can do only feebly or not at all. The distinctive *something* of the black parent therefore dominates a corresponding *nothing* of the white parent. White fowls, on the other hand, are not albinos. They have pigmented eyes. Accordingly they do not lack the power to form pigment, owing to the absence of some necessary ferment or pigment ingredient.

White guinea-pigs occur which are in a way comparable with white fowls. They look exactly like albinos, except that their eyes are black, whereas the eyes of the albino are pink. If such a black-eyed white guinea-pig is crossed with an albino of the sort shown in Fig. 15, the young produced will be black all over. Now this result shows that the black-eyed white animal *possesses* what is lacking in the albino as compared with the all-black animal. It would seem, therefore, that it lacks something different from what the albino lacks, and that a cross of the two supplies *both lacks*, the albino supplying what is wanting in the black-eyed white, and *vice versa*. Accordingly, wholly black offspring result from the crossing of the two white races.

But the case of white poultry is different from this, since white poultry lack *nothing* that is necessary to produce the complete black plumage. For when white fowls crossed with black ones produce *white* offspring, if these offspring are then bred with each other, they produce both white offspring and black ones in the ratio 3 to 1. White fowls, therefore, *are* able to produce the black condition. This ability is in the white individual held in abeyance, it is not exercised. Why, we do

not know. Some suppose it to be held in
check by an additional unit-character, an in-
hibiting factor, but we have no direct evidence
that such a factor exists. All that we are
warranted in saying at the present time is
that black and white in poultry represent *different* conditions of pigmentation, alternative
to each other in heredity. In crosses of the
two, white is ordinarily dominant over black,
but in crosses between certain strains of white
and black poultry this relationship is reversed,
as Bateson has shown.

In still other cases, a cross of white with
black fowls produces offspring which resemble
neither parent closely, but which are in reality
intermediate. They are known as blue or
Andalusian fowls. They manifest a dilute
condition of black, such as one might obtain
by mixing lampblack with flour; they are in
reality a fine mosaic of black with white.
Such a condition has thus far been obtained
only from a cross of black fowls with a peculiar strain of impure sooty whites. This strain
undoubtedly contains the mosaic pattern but
without sufficient black pigment to make it

plainly visible. A cross with a black race makes it visible. No one, however, has succeeded in "*fixing*" a blue race, that is, in obtaining a strain which would breed true.

When two blue individuals are bred together they produce black, blue, and white offspring in the ratio $1:2:1$. The blacks are homozygous, B B; the whites also are homozygous, W W, but the blues are invariably heterozygous, B W. Blue accordingly in this case is called a *heterozygous* character, one which is due to the presence in one zygote of two unlike unit-characters, which invariably segregate from each other at the genesis of gametes, but which jointly produce a different appearance from what either produces by itself. If a strain of Andalusian fowls should ever be secured which would breed true, it would have to come about by the association of black with white in a *non-segregating* relationship, so that *both* would be transmitted in the same gamete. That is, one would have to secure in the same gamete with white enough black pigment to bring out the latent mosaic pattern, and further, one would have to secure a homozygous race of fowls which formed such gametes.

Success would be most likely to attend the experiment if one selected always the sootiest whites obtained from blue parents, for blue results, as we have seen, from the association of *more black* with the white and in the pattern borne already by the white race.

A much-debated case of inheritance which involves this principle of unfixable heterozygous characters occurs among fancy mice, in the variety known as yellow. A wonderful series of color varieties exists among mice kept as pets, equalling or perhaps surpassing that known in the case of any other mammal. All these varieties appear to be derivatives of the common house-mouse, with which they cross readily. All are capable of explanation as unit-character variations from the condition of the house-mouse. Among all these varieties yellow is most peculiar in its behavior. In crosses it is dominant over all others, yet is itself absolutely unfixable.

If certain strains of yellow mice are crossed with black ones, the offspring produced are of two sorts equally numerous, yellow and black. From this result alone it is impossible to say which is the dominant character, but breeding tests of the offspring show that yellow is the dominant character. For the black offspring bred together produce only black offspring, but the yellows bred together produce both yellow offspring and black ones. The curious feature of the case is that when yellows are bred with each other no pure yellows, that is, homozygous ones, are obtained. Hundreds of yellow individuals have been tested, but the invariable result has been that they are found to be heterozygous; that is, they transmit yellow in *half* their gametes, but some other color in the remaining gametes — it may be black, or it may be brown, or it may be gray. The black, brown, or gray animals obtained by mating yellow with yellow mice never produce yellow offspring if mated with each other. This shows that they are genuine recessives and do not contain the yellow character, which is dominant.

Now ordinary heterozygous dominants, when mated with each other, produce three domi-

⌐ nant individuals to one recessive. Accordingly we should expect yellow mice, if, as stated, they are invariably heterozygous, to produce three yellow offspring to one of a different color, but curiously enough they do not. They produce *two* yellows (instead of the expected three) to every one of a different color. About the ratio there can be no reasonable doubt. It has been determined with great accuracy by my pupil, Mr. C. C. Little, who finds that in a total of over twelve hundred young produced by yellow parents almost exactly two-thirds are yellow. Instead of the regular Mendelian ratio 3 : 1, we have then in this case the peculiar ratio 2 : 1, and this requires explanation. The explanation of this ratio is to be found in the same circumstance as is the total absence of *pure* yellows. Pure yellow zygotes are indeed formed, but they perish for some unaccountable reason. For a yellow individual forms gametes of two sorts with equal frequency, viz. yellow and non-yellow (let us say black). For, if yellow individuals are mated with black ones, half the offspring are black, half yellow, as already stated.

If now yellow individuals are mated with
each other we expect three sorts of young to
be produced numerically, as 1 : 2 : 1, viz. 1 Y Y,
2 Y B, and 1 B B. But since observation shows
that only *two* combinations are formed which
contain yellow to one not containing yellow,
and since further all yellows which survive
are found to be heterozygous (Y B), it must
be that the expected Y Y individual either is
not produced or straightway perishes. As to
which of these two contingencies happens we
also have experimental evidence. Mr. Little
finds that yellow mice when mated to black
ones produce larger litters of young than when
they are mated to yellow ones. The average-
sized litter contains something like 5.5 young
when the mate is a black animal, but only 4.7
when it is a yellow animal. It is evident, then,
that about one young one out of a litter perishes when both parents are yellow, and this
undoubtedly is the missing yellow-yellow zygote. The yellows which are left are heterozygous yellow-black zygotes, and they are to
those that perish as 2 : 1. They are also to
the non-yellow zygotes as 2 : 1, the ratio ob-

served also among the surviving young of yellow by yellow parents.

This interpretation of the 2 : 1 ratio observed in this case is strongly supported by a similar case among plants, in which the evidence is even more complete. A so-called " golden " variety of snapdragon, one in which the foliage was yellow variegated with green, was found by the German botanist, Baur, to be unfixable, producing when self-pollinated fully green plants as well as golden ones, in the ratio 2 golden : 1 green. The green plants were found to breed true, that is, to be recessives, while the golden ones were invariably found to be heterozygous. Baur found, however, by germinating seeds of golden plants very carefully, that there were produced in addition to green plants and golden ones a few feeble seedlings entirely yellow, not variegated with green, as the golden plants are. These, for lack of assimilating organs (green chlorophyl), straightway perished. Clearly they were the missing pure yellow zygotes.

Some Mendelian characters, while not themselves heterozygous and so unfixable, are nevertheless produced only when two independently inherited factors are present together. A character of this sort does not itself conform with the simple Mendelian laws of inheritance, but its factors do. Herein lies the explanation of atavism or reversion, and the process by which reversionary characters may be fixed.

Atavism or reversion to an ancestral condition is a phenomenon to which Darwin repeatedly called attention. He realized that it is a phenomenon which general theories of heredity must account for. He supposed that the environment was chiefly responsible for the reappearance in a species of a lost ancestral condition, but that in certain cases the mere act of crossing may reawaken slumbering ancestral traits. Thus he noticed that when rabbits of various sorts are turned loose in a warren together, they tend to revert to the gray-coated condition of wild rabbits. And when pigeons are crossed in captivity they frequently revert to the plumage condition of the wild rock pigeon, *Columba livia.* In plants, too, Darwin recognized that crossing is a frequent cause of reversion. The explanation which he gave was the best that the knowledge of his time afforded, but it leaves much to be desired. This lack, however, has been completely supplied by the Mendelian principles. An illustration or two may now be cited.

When pure-bred black guinea-pigs are mated with red ones, only black offspring are as a rule obtained. The hairs of the offspring do indeed contain some red pigment, but the black pigment is so much darker that it largely obscures the red. In other words, black behaves as an ordinary Mendelian dominant. In the next generation black and red segregate in ordinary Mendelian fashion, and the young produced are in the usual proportions, three black to one red, or $1:1$ in back-crosses of the heterozygous black with red. All black races behave alike in crosses with the same red individual, but among red animals individual differences exist. Some, instead of behaving like Mendelian recessives, produce in crosses with a black race a third apparently new condition, but in reality a very old one, the agouti type of coat found in all wild guinea-pigs, as well as in wild rats, mice, squirrels, and other rodents. In this type of coat reddish yellow pigment alone is found in a conspicuous band near the tip of each hair, while the rest of the hair bears black pigment. The result is a brownish or grayish ticked or grizzled coat, inconspicuous, and hence protective in many natural situations.

Some red individuals produce the reversion in half of their young by black mates, some in all, and others, as we have seen, in none, this last condition being the commonest of the three. It is evident that the reversion is due to the introduction of a third factor, additional to simple red and simple black. It is evident further that this new third factor, which we will call A (agouti), has been introduced through the red parent, and that as regards this factor, A, some individuals are homozygous (AA) in character, others are heterozygous (transmit it in half their gametes only), while others lack it altogether. Further observations show that it is independent in its inheritance of both black and red; it is in fact an independent Mendelian character, which can become visible only in the presence of both black and red, because it is a mosaic of those two pigments. If the F_1 agouti individuals are bred together they produce in the next generation (F_2) three sorts of young, viz. agouti, black, and red, which are numerically as $9:3:4$. This evidently is a modification of the dihybrid Mendelian ratio $9:3:3:1$, resulting from the fact that the last two classes are superficially alike. They are red animals with and without the agouti factor respectively; but this agouti factor is invisible in the absence of black, so that both sorts of reds look alike. Together they number four in sixteen of the F_2 offspring.

Fig. 35 is intended to show how the independent factors behave in heredity. The black parent contributes the factor B, the red parent, R and A, so that the zygote, or new individual, contains the three factors necessary to the production of agouti. When the new individual forms gametes (sex-cells), these will be of four different kinds, for A is independent of B and of R and may pass out with either one in the reduction division which sepa-

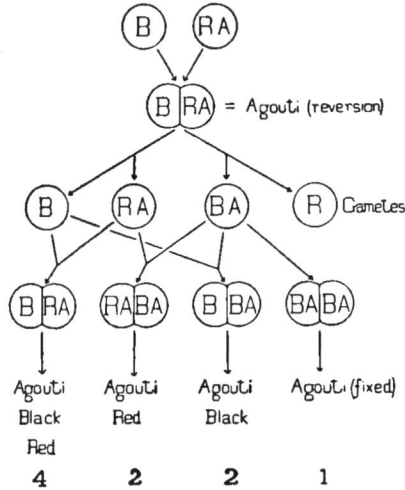

Fig. 35. — Diagram to show the gametic combination and re-combinations which occur in the production and fixation of an atavistic coat-character in guinea-pigs.

Row 1 shows the character of the gametes formed by the parents crossed; row 2 shows the character of the F_1 agouti individuals resulting from the cross; row 3 shows the two different sorts of gametic splittings which may occur in the production of gametes by the F_1 agoutis, and how four different kinds of gametes result; row 4 shows how among such gametes four different kinds of unions may occur that will produce agouti young. The BA·BA combination, it will be understood, could result only from the union of a BA gamete with another gamete of like constitution. Below each of the four combinations is indicated the kinds of young which an animal of that sort would produce if mated with an animal like itself. The numerals show the expected relative frequencies of the four sorts of combinations.

" rates B from R. That division accordingly may occur either so as to form gametes B and R A respectively, or what is equally probable, so as to produce gametes B A and R. Observation confirms this interpretation, for it is found that the reversionary agoutis do not breed true, but produce young of the three sorts, agouti, black, and red, as expected. We expect black individuals from unions of B with B, or of B with R; we expect red individuals from unions of R with R or with R A, and from unions of R A with R A; we expect agoutis to be produced by any gametic union which brings together the three factors B, R, and A. There are six chances in sixteen for the occurrence of such a union, when the reversionary agoutis are bred together. In fact, however, agoutis are produced much oftener. Approximately nine out of sixteen of the young have been found to be agoutis. The unexpected excess of agoutis in our experiments was fully explained when these second-generation agoutis were tested individually. It was then found that they are of four sorts as regards breeding capacity. The first sort produces the three kinds of young, agouti, black, and red; the second sort produces only agouti young and red young; the third sort produces only agouti young and black young. The fourth sort produces only agouti young, i. e. represents the fully fixed agouti type, the completely recovered wild type.

In the chart (Fig. 35) are indicated certain gametic unions which would lead to the production of these four classes of agoutis. The probable frequencies of their occurrence on the basis of chance are $4 : 2 : 2 : 1$.

Experiment made it clear that R as an independent gametic factor is not necessary to the production of the agouti character, as was at first thought to be the case, but that any gametic union which includes both B and A will produce an agouti individual whether R is or is not present. Yet a microscopic examination of the agouti hair shows that red pigment is present in a distinct band near the hair-tip. As a matter of fact all black individuals, even when they breed true, probably form some red pigment along with the black, but its presence is overlooked when the more opaque black is distributed throughout the whole length of the hair. When, however, black is excluded from the hair-tip, the red then becomes visible as the agouti marking; elsewhere the hair appears black. Red, then, we may assume, is always present with black in sufficient quantities to produce the agouti marking if the factor A is present (absence of black from the hair-tip). This explains why blacks never give the reversion in any sort of cross, but it is always brought about through the agency of the red parent. If a black individual contained the factor A, it would no longer be a black individual, but an agouti one.

The existence of a third factor, A, in cases of reversion in coat-character among rodents was long overlooked merely because it does not represent a distinct pigment or set of pigments, but consists in a particular kind of pigment distribution on the individual hairs. The agouti hair is due to a definite cycle of activity of the hair follicle in forming its pigments, — first black, then red, then black; the wholly black hair is due to a continuous process of pigment formation without alternation in the character of the pigments produced.

In rabbits as well as in guinea-pigs reversion to the original wild type, in this case gray, may be obtained by crossing a black animal with a yellow one. In guinea-pigs the yellow (or red) animal which will yield this result cannot be distinguished in appearance from one which will not; but in rabbits the yellow animal which will give reversion has a white belly and tail, while the one which will not give reversion is not so distinguished.

We now know what is implied in the fixation of a heterozygous character obtained by crossing. When A and B are crossed we obtain a third condition, C. C is due either to the simple coexistence of A with B, or to the coexistence with them of a third factor introduced with one or the other. In either case fixation will consist in getting into the *gamete* all the factors necessary to the production of C. In the first supposed case the zygote is A·B and the resultant is equivalent to C. Fixation will consist in getting a zygote of the formula AB · AB. In the second supposed

case the zygote produced is either A·CB or
AC·B; fixation will consist in obtaining a zygote ACB·ACB; every gamete formed will
then contain the three factors A, C, and B.

BIBLIOGRAPHY

BATESON, W.
1909. "Mendel's Principles of Heredity," 393 pp., illustrated. University Press, Cambridge; also G. P. Put-
nam's Sons, N. Y. [Contains translation of Mendel's
original papers.]

BAUR, E.
1907. "Untersuchungen über die Erblichkeitsverhältnisse
einer nur in Bastardform lebensfähigen Sippe von An-
tirrhinum majus." *Ber. d. Deutsch Bot. Gesellsch.*, 25,
p. 442.

CASTLE, W. E.
1907. "On a Case of Reversion Induced by Cross-Breeding and its Fixation." *Science, N. S.*, 25, pp. 151–153.
1907. "The Production and Fixation of New Breeds."
Proc. Amer. Breeders' Ass'n, 3, pp. 34–41.

CASTLE, W. E. and LITTLE, C. C.
1910. "On a Modified Mendelian Ratio Among Yellow
Mice. " *Science, N. S.*, 32, pp. 868–870.

CUÉNOT, L.
1908. "Sur quelques anomalies apparentes des proportions Mendeliénnes." *Arch. Zoöl. Exper.* (4), Notes
et Revue, p. vii.

DAVENPORT, C. B.
1906. "Inheritance in Poultry." *Carnegie Institution
of Washington, Publication No. 52*, 104 pp., 17 pl.
1909. "Inheritance of Characteristics in Domestic Fowl."
*Carnegie Institution of Washington, Publication No.
121*, 100 pp., 12 pl.

CHAPTER V

OUR knowledge of Mendelian phenomena is most complete in the case of color-inheritance. We find that the flower-colors of plants and the coat-colors of mammals are alike complex, and that what seem at first sight simple results may really depend on several independent factors acting conjointly. By analysis of such complex cases we are able to gain some idea of what the probable course of evolution has been in the production of the color varieties found among cultivated plants and domesticated animals.

Thus among rodents (mice, rabbits, guinea-pigs) the coat is grayish, consisting of black, brown, and yellow pigments mingled together on the same individual hair in a pattern of greater or less complexity.

72

The simplest variation from this ancestral type of coloration is albinism, a wholly unpigmented condition in which the eyes are pink. This is due to the loss of the capacity to form pigment. Albinism is recessive in crosses. We explain it by assuming that something necessary for color production is wanting in the albino, and call that something the color-factor C, without necessarily making any assumption as to its nature. Another common variation is the loss of the pattern-factor of the individual hair, the agouti or A factor. An account of the discovery of this factor was given in the last chapter. In consequence of the loss of this factor the pigments become mingled together without order, and the result is a uniform black, the denser pigment hiding the others.

A third variation is the loss of the capacity to form *black* pigment (factor B), only brown and yellow pigments being left. Thus arise brown and cinnamon varieties. Through these three independent loss-variations there arise eight different color-varieties as follows:

Gray (or agouti)	= C B Br A;	Cinnamon	= C Br A
Black	= C B Br;	Brown	= C Br
Albino (1)	= B Br A;	Albino (3)	= Br A
Albino (2)	= B Br;	Albino (4)	= Br

Proof of the correctness of this interpretation may be obtained from crosses. Suppose the four kinds of albinos described be crossed with the same colored variety, brown; albino 1 will produce gray offspring, albino 2 will produce black ones, albino 3 will produce cinnamon ones, and albino 4 will produce brown ones. The cross with albino 1 brings together all the four factors entering into the production of gray, viz. C, B, Br, and A, hence the young are gray. The cross with albino 2 brings together the factors C, B, and Br only. The result is black. The cross with albino 3 brings together the factors C, Br, and A; result, a cinnamon animal. The cross with albino 4 brings together no factors except C and Br; result, a brown animal.

Thus far we have considered merely variations which arise by loss of one or more of the three unit-characters, A, B, and C. We may now consider variations which arise by modification without loss of these same factors.

Yellow varieties owe their origin to a reduction in the amount of black or brown pigment in the fur, and to a corresponding increase in the amount of yellow. In some yellow animals, such as the sooty yellow rabbit, black and brown pigments are not wholly lacking in the fur, but are only greatly reduced in amount. They always persist in the eye. In other yellow animals, mice for example, the black or brown pigments are wholly absent from the fur, and they may also be greatly reduced in amount in the eye, as in the variety known as pink-eyed yellow, but in no yellow animal, so far as I am aware, is the production of black and of brown pigments wholly suppressed.

In any mammal which possesses yellow varieties we can produce by suitable crosses as many different varieties of yellows as there are of gray, black, cinnamon, and brown varieties combined. For example, in mice, yellow individuals of which, as was shown in the last chapter, are invariably heterozygous and produce some other variety than yellow, even when mated with yellows, we can recognize the following varieties distinct in breeding capacity, though all looking very similar.

1. Yellows which produce yellow young and gray ones;
2. Yellows which produce yellow young and black ones;
3. Yellows which produce yellow young and cinnamons
4. Yellows which produce yellow young and brown ones.

Albino varieties occur which correspond with each of these yellow varieties, viz. (1) albinos which if crossed with brown will produce yellow young and gray ones; (2) albinos which crossed with brown produce yellow young and black ones; (3) albinos which crossed with brown produce yellow young and cinnamon ones; and (4) albinos which crossed with brown produce yellow young and brown ones. Such albinos, of course, differ from the corresponding yellow varieties merely by the general color factor C, which the albino lacks. If this is added by a cross, they produce the same visible result as the corresponding yellow variety in the same cross.

In addition to the modification which produces yellow varieties, we can recognize several other modified conditions of the unit-characters A, B, C, and Br, which modifications produce whole series of color varieties. For a modified condition of a single unit-character is capable of producing as many new varieties as there are possible combinations of the modified character with other unit characters.

One who attends a poultry-show cannot fail to be impressed with the great number of color varieties among poultry. Let him first observe these among fowls of common size, and if he then visits the bantam section he will find them all duplicated in miniature among the bantams. If a new color variety is brought out, it is only a short time until it finds its place among the bantams as well as among fowls of common size. The dwarf size of the bantam is clearly due to a modified condition of one or more unit-characters capable of combinations with as many different kinds of coloration as occur among poultry. The various combinations are of course brought about by crossing, and two generations suffice theoretically for securing them.

In mice, if one possessed only the albino variety last described, — the one which corresponds with the brown-eyed yellow variety, — he could easily produce within six months every one of the various color varieties which have been mentioned. All he would have to do would be to catch some wild mice and cross these with his albinos. The immediate offspring produced by the cross might seem unpromising; they would either be gray, exactly like wild mice, or else yellow. But if our breeder possessed the faith to breed a second generation from these animals, he would be rewarded by seeing all the color varieties which I have described put in an appearance, viz. yellows with black eyes, and yellows with brown eyes, blacks, browns, cinnamons, and grays, and albinos corresponding in character with each colored variety except for the lack of the color-factor C.

It may be of interest to consider how some additional color varieties of mice have arisen, for of all mammals bred in captivity the mouse is probably richest in color varieties. In one series of these the capacity to form black or brown pigments is greatly weakened, so that the coat is less heavily pigmented and the eye is *almost* wholly unpigmented, and looks pink, due to the red color of the blood in the eye. This series we may call the pink-eyed series. All the common color varieties occur in a pink-eyed as well as in a dark-eyed series. Thus there are pink-eyed grays, pink-eyed blacks, pink-eyed cinnamons, pink-eyed browns, and pink-eyed yellows, as well as albinos which transmit the pink-eyed condition in crosses.

Given a single pink-eyed individual in any one of these varieties, all the others may be produced from it by suitable crosses. Thus a pink-eyed gray crossed with brown produces in F_1 reversion to the condition of the wild house-mouse, but in F_2 (that is, among the grandchildren) occur eight varieties, — four dark-eyed and four pink-eyed. Gray, black, cinnamon, and brown occur, both in dark-eyed and in pink-eyed individuals, the latter being also far lighter in color than the dark-eyed varieties. The pink-eyed condition is therefore in mice a unit-character modification of the pigmentation, independent of any of the pigment factors previously mentioned, since it can be transferred by crosses from association with one of these to association with another. It may also be transmitted equally well through colored and through albino individuals, though it produces a visible effect only in colored individuals.

Another unit-character modification of the pigmentation seen in mice produces a series of dilute or pale pigmented varieties, but different in character from the pink-eyed series, since their eyes may be dark, not pink. The pale modification of gray is known to fanciers as " blue-gray," that of black is known as " blue," and that of brown is known as " silver fawn." The pale quality is interchangeable between black, brown, and yellow pigmentation, so that if one has a pale gray variety he may by crosses obtain also pale black, pale cinnamon, pale brown, and pale yellow varieties. Or if one starts with pale yellow, he may by crosses with a perfectly

" wild mouse obtain also pale gray, pale black, pale cinnamon, and pale brown varieties, all within two generations from the cross.

Now the pale modification is distinct from the pink-eyed modification, and independent of it in transmission. Accordingly, it is possible to have the two modifications combined in the same race. Thus arises a series of pale pink-eyed grays, blacks, cinnamons, browns, and yellows. Since paleness is in crosses recessive to intense pigmentation, and pink eyes are recessive to dark ones, it follows that a variety which is both pale and pink-eyed will breed true to those characteristics without fixation.

The lightest colored of the pale pink-eyed varieties develop very little pigment indeed, yet the modifications to which they are due are wholly different in nature from the albino variation, as a very simple experiment will show. Cross together an albino of variety (1), page 74, — which is a snow-white animal with pink eyes, — and a pale pink-eyed, brown animal, whose coat is pale straw color, and whose eyes, like those of the albino, are pink. Although both parents are pink-eyed, and one develops no pigment whatever in its fur, while the other develops very little, nevertheless the offspring are as dark as the darkest wild mice, eyes, fur, and all. They look just like common house-mice. This result shows that the albino variation is something very different in nature from the modifications found in the pink-eyed brown parent, since each parent contains those constituents of the wild gray coat which the other parent lacks.

I can think of no more instructive laboratory experiment illustrative of Mendelian inheritance than to follow through two generations the cross just described, and to analyze critically the results obtained. One who does this can never be sceptical about the value of crossing as an agency in the production of new varieties. For in the second generation from the cross he will obtain (1) ordinary gray, black, cinnamon, and brown varieties; (2) *pale* gray, black, cinnamon, and brown varieties; (3) *pink-eyed* gray, black, cinnamon, and brown varieties; (4) *pink-eyed and pale* gray, black, cinnamon, and brown varieties; and lastly, albinos, which, if he has the patience to test them one by one, will prove to be of sixteen different homozygous kinds, to say nothing of the much more numerous heterozygous sorts.

No mention has thus far been made of spotted races, in which a unit-character modification has occurred which results in a distribution of pigment to part of the coat only, the remainder being unpigmented. Although this modification apparently regulates the distribution of pigment over the body, it is independent of the general color factor C, since it is transmitted through albinos, which by hypothesis lack C.

Spotting is also independent of all the other unit-character modifications which have been described. Consequently we have in mice four different series of spotted varieties, — the intense spotted, the dilute spotted, the intense pink-eyed spotted, and the dilute pink-eyed spotted. In each of these series are gray, black, cinnamon, brown, and yellow individuals, making a total of twenty spotted sorts, all of which may be obtained from crossing a single pair of properly selected parents, such, for example, as an albino and a wild house-mouse of the kind every barn contains.

The color variations of guinea-pigs are similar to those of mice; the same series of unit character changes has produced them with one exception. The pink-eyed modification is wanting in guinea-pigs. We are therefore limited here to the intense series, the pale series, the intense spotted series, and the pale spotted series. In each of these occur gray (or agouti) individuals, black ones, cinnamon ones, and brown ones.

The parallelism between the color variations in guinea-pigs and in mice received an interesting demonstration in a particular case. The brown pigmented series in mice has been known for some time, but in guinea-pigs the brown variety is of comparatively recent origin, and the cinnamon variety was wholly unknown until some three years ago. After an analysis had been made in terms of unit-characters of the color varieties of the mouse, it became clear that if the color variation of guinea-pigs followed a like course, a then unknown variety of guinea-pig, cinnamon, should be capable of production by crossing an agouti animal with a brown one. In 1907 a statement of the scientific expectation in the case was published, and a few months later I had the satisfaction of announcing its fulfillment in the second generation (F_2) from the cross in question.

The experiment progressed as follows: The parents were an agouti and a black, their F_1 offspring were agoutis in character; but the F_2 offspring were of four sorts, — *agouti, black, cinnamon,* and *brown.* The cross thus produced two varieties new to the experiment, viz. black and cinnamon, the latter being a variety at that time new among guinea-pigs.

The subsequent behavior, too, of the newly produced cinnamon variety is in harmony with expectation based on Mendelian principles. The cinnamon variety has not produced agouti or black individuals, which from the formulæ it will be seen it may not be expected to produce, since it lacks the factor B. But it has in some cases produced brown individuals, as it clearly could in case both parents to a mating were heterozygous (single) in factor A.

On the whole the evidence seems very clear that the numerous color varieties of animals kept in captivity arise chiefly from loss or modification of Mendelian unit-characters. Loss of a unit-character might easily come about by an irregular cell-division in which the material basis of a character failed to split, as normally. The consequence would be that the character in question would be transmitted by one only of the two cell-products produced. The cell lacking a character might be the starting-point of a race lacking the character, as of a black race, derived from a gray one. On the other hand a modified condition of a unit-character might possibly result from *unequal* division of the material basis of a character, so that one of the cell-products would transmit the character in weakened intensity, the other in increased intensity.

BIBLIOGRAPHY

CASTLE, W. E.
  1907. "Color Varieties of the Rabbit and of Other Rodents: Their Origin and Inheritance." *Science, N. S.,* 26, pp. 287–291.
  1908. "A New Color Variety of the Guinea-pig." *Science, N. S.,* 28, pp. 250–252.
  1909. "Studies of Inheritance in Rabbits." *Carnegie Institution of Washington, Publication No. 114,* 70 pp., 4 pl.

CHAPTER VI

EVOLUTION OF NEW RACES BY VARIATIONS IN THE
POTENCY OF CHARACTERS

IN the last chapter we discussed the color
variations of mammals, and we concluded
that these result largely from the loss or
modification of some half-dozen independent
Mendelian unit-characters. As to the material
basis of these unit-characters some interesting
evidence has recently been collected by Riddle.
Melanin pigment has been for some time known
to be formed by oxidation. A variety of organic compounds may undergo oxidation into
melanin pigments ranging in intensity from
light yellow to black; the greater the oxidation, the darker the product. But it is not
certain, as assumed by Riddle, that the chemical *method* of oxidation is the same in all cases
or that the substance to be oxidized is the same.
The results obtained from breeding experiments

show that the capacity to form pigment of all sorts may be lost by a single variation, which we have called loss of the color factor, C. We do not know whether it consists in the loss of a substance capable of oxidation, or of the power to take some indispensable first step in the process of oxidation, perhaps due to loss of an enzyme; but we do know that when this particular variation has occurred, the power to produce other than albino individuals cannot be recovered by any known means except a cross with colored animals. We know also that the capacity to form specific kinds of pigment (yellow, brown, or black) is independent of the general color-factor, C; for albinos may transmit those specific powers without themselves being able to form any kind of pigment at all, i. e. without possessing C. Any animal which forms pigment of one of the higher grades has the capacity apparently to form pigment also of the lower grades. Thus a black animal can form also brown and yellow pigment granules. Brown (chocolate) animals, however, lack the capacity to form black pigment. The oxidation, it would seem, can in this case be carried no further than the brown stage, because of the lack of some oxidizing agency necessary to the last stage in pigment production. The production of yellow is probably a first or early step in the oxidation process preliminary to the production of brown or black, yet all yellow animals, so far as known, are able to take the further steps; they retain the capacity to form either brown or black pigment to some extent, if only in the eye.

The variations thus far described are what De Vries has called retrogressive, i. e. due to loss or modification. A much rarer sort of variation has been called by De Vries progressive, i. e. due to gain, acquisition of some character not before possessed by the race. I can call to mind very few cases which certainly fall in this category. One which it would seem must belong here is the rough or rosetted condition of the hair in guinea-pigs, a variation similar in nature to the reversed plumage of birds, seen, for example, in the Jacobin pigeon. The rough coat of guinea-pigs is surely not an ancestral condition, yet it behaves as a dominant character in crosses. It can scarcely be explained by loss; the only alternative is to consider it an acquisition, unless we choose to consider it a modification of the normal condition.

Aside from the sorts of variations already discussed, which consisted either in the loss or modification of existing unit-characters or in the gain of new ones, we must also recognize, as a cause of permanent and heritable variation, changes in the *potency* of unit-characters, i. e. their tendency to dominate in crosses.

When a gamete containing a particular unit-character unites with a gamete not containing it, the zygote formed will ordinarily show the character in question fully developed. This result following Mendel's terminology we call dominance. But dominance is frequently imperfect and may even be reversed. The zygote in which a character is doubly represented frequently develops the character more fully than the zygote in which it is represented but once. If a black guinea-pig is crossed with a yellow one the offspring are black, but oftentimes of a slightly yellowish shade. Likewise if black

is crossed with brown, the crossbreds are apt
to develop in their coats more brown pigment
granules than do homozygous or pure blacks.
Nevertheless, we have no reason to question the
entire purity of the gametes, both dominant
and recessive, formed by such cross-bred black
animals. It is the dominance, not the segregation, which is imperfect.

In other cases still the dominance may be
entirely reversed in character, owing to variation in the potency of a unit-character. Thus
in most rodents the gray or agouti pattern-
factor of the hair, A, is dominant. A cross of
black with homozygous gray, in rats, mice, or
rabbits, produces only gray offspring, which
in F_2 produce three grays to one black. But
the so-called black rat, *Mus rattus*, a species
distinct from the one which has given rise to
the varieties kept in captivity, behaves in a
different way, as shown by Morgan ('09).
When crossed with its gray variety, the roof
rat, *Mus alexandrinus*, it produces only *black*
offspring, and in F_2, three blacks to one gray.
If we suppose the gray coat in this case to
be due to the same factor as in other rodents,

we must assign to it a different potency, or power of dominance, so that it produces a visible effect only when doubly represented in the zygote.

In guinea-pigs, rabbits, and mice we have seen that the presence together in the same zygote of two factors, A and B, in any combination whatever, produces the gray or agouti coat. The two factors are A, the agouti or gray marking of the hair, and B, black pigment in the fur. If A is lacking, the coat is black; if B is lacking, it is brown, cinnamon, or yellow. If both are lacking, it is either brown or yellow. But if both are present, the wild or agouti type is produced. So far as the production of the agouti coat is concerned, it makes no difference whether either factor is singly or doubly represented in the zygote. Each factor has potency enough to produce the full effect either in a single or in a double dose. Accordingly, as we noticed in an earlier chapter, we can distinguish by their breeding capacity, though not by their looks, four types of agouti guinea-pigs or gray rabbits, viz.:

1. A A B B, which breeds true, since it forms gametes all A B;
2. A B B, which produces agouti young and black ones in the ratio 3 : 1, since it forms gametes A B and B;
3. A A B, which produces agouti young and yellow ones in the ratio 3 : 1, since it forms gametes A B and A;
4. A B, which produces agouti, black, and yellow young in the ratio 9 : 3 : 4. For the gametes formed by this sort are of four kinds, A B, A, B, and neither A nor B.

Now in rats we have no evidence that the factor B has ever been lost, a matter to which we shall presently return; but the agouti factor is apparently frequently wanting in ordinary rats, which are then black. For ordinary rats, then, the known combinations of A and B seem to be three, viz.:

A A B B = the pure gray (wild type);
A B B = heterozygous gray, which produces offspring 3 gray : 1 black. This type is obtained by crossing black with wild gray;
B B = pure black.

Now in *Mus rattus,* as we have seen, the middle or heterozygous type is *black*, not gray in appearance, but it produces both the gray and the black types. So the same gametic formulæ will account for both sets of facts, if we suppose merely that the *potency* of A is different in the two cases. In ordinary rats (*Mus norvegicus*) A produces the gray coat in a single dose; but in *Mus rattus* its potency is less, two doses are required to produce the gray coat. I am unable to frame any hypothesis other than this which will account for the reversal of dominance in one case as compared with the other.

Yellow color in mammals affords another illustration of this same thing, — reversal of dominance. Black and brown are in most mammals dominant over yellow in crosses, but in mice the reverse is true. The differential factor between black and yellow, if it is the same in mice as in other rodents, must be in one case potent enough to show itself if singly represented in the zygote, whereas in the other case it produces no visible effect unless doubly represented in the zygote. Yellow certainly seems to be a retrogressive variation from gray, black, or brown. The pigment granules remain in a *lower* oxidation stage in yellow than in black or brown. We suppose that in the yellow animal something is wanting which makes that further oxidation possible. This hypothesis would fully account for the observed recessive nature of yellow in the case of all mammals except mice. But here the capacity to form black or brown pigment is regularly present in the yellow individual but is held in check. We may suppose, therefore, that the differential factor, that which converts yellow into brown or black, must in this case be *doubly* represented in the zygote in order to produce brown or black fur, whereas in most mammals a single dose is effective. Accordingly, if the unmodified black or brown factor is represented only *once* in the zygote, and the yellow modification is represented once, the latter will show, since the former is singly ineffective. The animal accordingly is a heterozygous yellow, capable of producing also black or brown offspring. But mice are peculiar in that they cannot exist in the doubly deficient condition of a pure yellow zygote, consequently *all* yellow mice are heterozygous dominants,

95

whereas other yellow mammals are homozygous recessives.

In connection with this same case may possibly be found the explanation of the complete absence of the yellow variation in rats. In nearly all mammals kept in captivity yellow as well as black varieties occur; this is true of horses, cattle, swine, dogs, cats, rabbits, guinea-pigs, and mice. In rats, however, a yellow variety is unknown. We know that rats are able to form yellow pigment, for all wild rats do form yellow pigment in their agouti fur, yet singularly enough no *all-yellow* rat has ever been observed, so far as we have any record, either wild or in captivity. A rat of this sort would command a high price at the hands of any fancier. Suppose the variation did occur in a single gamete. If, as in most mammals, it behaved as a recessive in crosses, it would not become visible, and might be carried along for untold generations without ever becoming visible unless two yellow gametes met. But if, as in mice, the yellow-yellow combination when formed quickly perished, then the character might never become visible. So the yellow variation may have occurred many times in rats, as it has in so many other mammals, but failed to become visible simply because it has the same potency as in most mammals, but is subject to the same *physiological limitations* as in mice, so that it cannot exist in a homozygous state. In that case the only evidence of its existence in a race would lie in a slightly diminished fecundity under inbreeding, as is found to be the case in yellow mice.

Such sharply contrasted variations in the potency of characters as we have been discussing are evidently of prime importance in evolution, making all the difference between a dominant and a recessive condition of a character, or between the occurrence and the permanent suppression of a particular variation. The character which is potent enough to show itself in a single dose will behave as a dominant character in crosses. We might call it *unipotent*. That which must be present in a double dose to produce a visible result will behave as a recessive character in crosses. We might call it *semi-potent*. It is not impossible that the *same character* may as regards domi-

97

nance behave in different ways under different circumstances, at one time dominating completely, at another only feebly, and at other times not at all.

Undoubtedly the chief condition affecting dominance is the nature of the gamete with which a union is made in fertilization. In 1905 (*Carnegie Inst. Publ. No. 23*) I described a case in which a particular guinea-pig (male 2002, shown in Fig. 32) having a rough or rosetted coat gave a varying result in crosses. In crosses with most smooth animals his rough character dominated completely (see Fig. 24, which shows a son of the male 2002 by a smooth mother), but with one particular smooth animal the dominance was very imperfect in all the young (Fig. 36), while with a second it was imperfect in half the young. The conclusion was drawn that gametes vary in potency, and that parents, too, differ as regards the potency of the gametes which they produce, some individuals producing gametes all of which are relatively potent, others producing gametes only half of which are potent, while still others produce gametes none of which are potent.

Relative potency would, therefore, seem to be a character inherited in Mendelian fashion.[1]

Observations of Coutagne on silk-moths may be cited in support of this idea. Coutagne made crosses between races of silk-moths dif fering in cocoon color, viz. between a race which spun yellow cocoons and another one which spun white cocoons. He found that *some* of the F_1 offspring spun yellow cocoons, others white ones. The F_1 yellow cocooned animals when bred together produced F_2 progeny which spun some yellow, others white cocoons, the two sorts being as $3:1$. In other words, yel- low in such cases behaved consistently as a *dominant* character. And the white-cocooned F_1 moths produced in F_2 cocoons of both colors, but in this case the white cocoons were to the yellow ones as $3:1$. In other words, when yellow behaved as a dominant in F_1 it behaved as a dominant also in F_2; and the same was true of white. Each retained throughout the two generations the *relative potency* with which

[1] It is of course possible to interpret such a case as due to the separate inheritance of a factor which inhibits the development of the character, but it is doubtful whether this line of explana- tion can be successfully applied to cases presently to be described.

99

it started. C. B. Davenport has also produced
much evidence favoring the idea of varying
potency of characters in recent papers based
on his extensive studies on poultry.

The case which I described in 1905 was one
in which unusual potency seemed to inhere in
the gametes of a recessive individual, — one
which apparently did not possess the character
whose dominance was affected. But there occur
also cases in which the varying gametic potency
is associated directly with the character affected. One such I was able to describe in
1906, — that of an extra toe in guinea-pigs. It
was found while building up a polydactylous
race by selection and crossing it with other
races that individuals varied in the potency
which the character had in their gametes. In
general the better developed the character was
in an individual the more strongly was it transmitted, i. e. the larger was the proportion of
polydactylous individuals produced in crosses.
In no case, however, was this a recognizable
Mendelian proportion, though both dominance
and segregation seemed to be taking place.
Variation in potency was, however, unmistak-

FIG 36

FIG 37

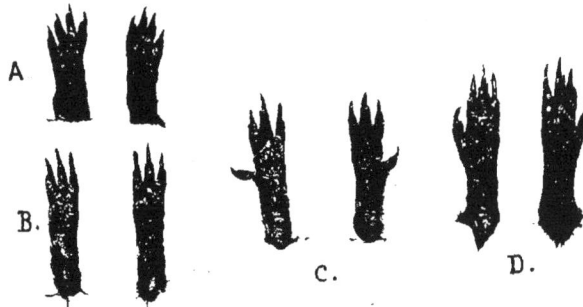

A

B.

C.

D.

FIG 38

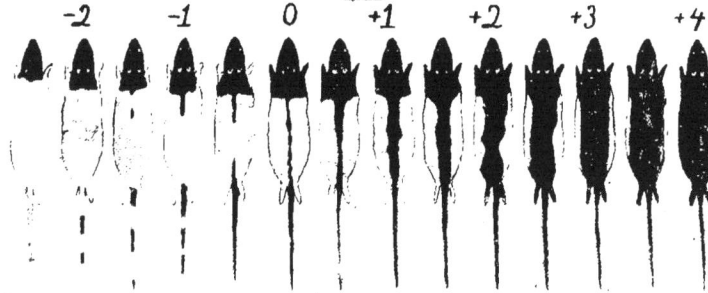

-2   -1   0   +1   +2   +3   +4

FIG 39

FIG. 36. — An imperfectly rough guinea-pig. Produced by mating the guinea-pig, shown in Fig. 32, with a particular smooth animal; female, 2005.

FIG. 37. — A silvered guinea-pig. One in whose coat occur white hairs interspersed with pigmented ones. The amount of the silver-

able and was transmitted from generation to generation.[1] See Fig. 36.

It is an important question whether potency is a property of the unit-character or of the gamete, i. e. whether it affects all the characters transmitted by a gamete or only a particular one. Practical breeders as a rule favor the idea of gametic rather than of unit-character potency, but this is probably due to a failure to discriminate between the two. They designate as " prepotent " an individual supposed to impress *all* its characters upon the offspring, but it is very doubtful whether such individuals exist. It is easy to mistake for an animal potent in *all* respects one which is potent in one or two important respects only, especially if the observer is unaware, as every one has been until quite recently, that one character is independent of another in transmission.

Conditions other than the character of the gametes themselves may determine the extent

[1] An alternative explanation is possible, viz. that the development of the fourth toe depends upon the inheritance of several independent factors, and that the more of these there are present, the better will the structure be developed. The correctness of such an interpretation must be tested by further investigations.

to which a character develops in the zygote, i. e. the completeness or incompleteness of its dominance in a particular case. For example, in salamanders, which apparently, like mammals, form skin-pigments of different sorts, such as yellow, brown, and black, Tornier has found that by feeding one may control the proportions in which chromatophores of the several sorts are formed in the skin. Abundant feeding causes preponderance of pigment of one sort, scanty feeding causes preponderance of pigment of another sort. Here external conditions determine the degree of development of characters. In other cases internal conditions may exercise a controlling influence. Thus in cattle the capacity to develop horns is a semi-potent unit-character, behaving as a recessive in crosses, heterozygotes developing only '' scurs,'' that is, mere thickenings of the skin, or else no trace of horns at all. In sheep, moreover, horns are more strongly developed in males than in females, the presence of the male sex-gland in the body, or rather probably some substance given off into the blood from the sex-gland, favoring growth of the horns.

In merino sheep the male has well-developed horns but the female is hornless; yet if the male is castrated early in life no horns are formed.

When a breed of sheep horned in both sexes, such as the Dorset, is crossed with one hornless in both sexes, such as the Shropshire, horns are borne by the male but not by the female offspring. Both sexes, however, are heterozygous in horns, as is shown by their breeding capacity. For in F_2 occur both horned and hornless individuals in both sexes. The hornless males and the horned females prove to be homozygous, but the horned males and the hornless females may be either heterozygous or homozygous. Accordingly the character, horns, behaves consistently as a dominant character in one sex, but as a recessive in the other. Further, the presence of the male sexgland in the heterozygote raises the potency of the character, horns, from semi-potent to unipotent, as the result of castration shows.

It is impossible to be certain that in a hornless race the character horns has been wholly lost. It may merely have fallen so low in potency that under ordinary conditions it pro-

103

duces no visible structures. The occasional occurrence of an imperfectly horned animal as a sport within a hornless race need not, then, occasion surprise. It would be a variation of the same sort as the extra toe in guinea-pigs (see Fig. 38), which, from a single sport, was built up by selection into a well-established race within a very few generations. This character, seemingly lost from the germ-plasm for an indefinite period, had perhaps merely fallen so low in potency that it no longer produced the fourth toe on the hind foot, though this was still present on the front foot. In the variant observed, the first polydactylus guinea-pig of my stock, the toe was imperfectly developed on one hind foot, doubtless as the result of an unusually potent condition of the character in one of the gametes which produced the individual. This manifestation of the character, though feeble, was sufficient to afford a guide for selection of those individuals which formed the most potent gametes, and so a polydactylous race was formed by selection and inbreeding.

Great as has been the contribution of Mendelian principles to our knowledge of heredity, they do not reduce the whole art of breeding to the production of new combinations of unit characters through crossing. Selection is required also, not merely among different combinations of unit-characters, but also among individuals representing the same combinations selection is required of those possessing the desired characters in *greatest potency*. The further rôle of selection in evolution we shall need to consider in a subsequent chapter.

BIBLIOGRAPHY

CASTLE, W. E., and LITTLE, C. C.
  1909. "The Peculiar Inheritance of Pink Eyes Among Colored Mice." *Science, N. S.*, 30, pp. 312–314.

COUTAGNE, G.
  1902. "Recherches expérimentales sur l'hérédité chez les vers-a-soie." *Bull. Sci.*, 37, pp. 1–194, 9 pl.

MORGAN, T. H.
  1909. "Breeding Experiments with Rats." *American Naturalist*, 43, pp. 182–185.

DE VRIES, H.
  1901–03. "Die Mutationstheorie." Leipzig, Veit and Co.

See also the Bibliographies to Chapters III., IV., and V.

CHAPTER VII

IF, as suggested in the last chapter, the potency of a character in crosses may be modified by selection, why may not the character itself be modified by selection, or are not the two things perhaps identical, viz. modification of the potency of a character and modification of the character itself? Darwin firmly believed that the characters of organisms can be modified by selection, and he made this the foundation stone of his theory of evolution. De Vries and Johannsen, however, have taught us a different doctrine, maintaining that selection is able to affect characters in superficial and transitory ways only, that the slight variations in characters which we see everywhere among organisms have no evolutionary significance or permanent value; that they come and go like the wavelets on the ocean beach, but have no more relation to evolution than the waves have to the tides. The brilliancy of the Mutation theory of De Vries, coupled with his great service to biology in rediscovering the Mendelian laws, has somewhat dazzled our eyes and led us, I think, to accept too readily his views concerning the efficacy of selection also. Ten years' continuous work in selection convinces me that much can be accomplished by this means quite apart from the process of mutation. The work of De Vries himself argues strongly in favor of this idea. To be sure, his interpretation of it is adverse to selection, and has seemed to most of us at times overwhelmingly convincing; but from his interpretation we may fairly appeal to the record of the work itself, and with this compare the record of our own work.

One of the most extensive selection experiments conducted by De Vries was made on the common buttercup, *Ranunculus bulbosus,* which occurs as a weed in pastures and meadows in this country as well as in Europe. It has, as is known, regular 5-petaled flowers. An examination of 717 flowers in the field made by De Vries in 1887 showed the rather frequent occurrence of 6 and 7 petaled flowers also, the average number of petals in the entire collection being 5.13. De Vries set himself the task to see if the proportion of many petaled flowers could be increased or the number of petals to a flower be further increased. In both these respects he succeeded surprisingly well. As a result of five successive selections the average number of petals was raised from 5.6 to 8.6, the upper limit of variation from 8 to 31, and the mode (or commonest condition) from 5 to 9. Singularly enough De Vries concludes, in accordance with general ideas which he had adopted, that selection had in this case done practically all that it could accomplish, that further selection, while it might advance the average somewhat farther, would have no permanent effect in modifying the type. This belief seems to have rested on considerations such as these. De Vries had found, as had others, that variations which are heritable have their origin in the germ-cells only. He recognized that the tendency to produce double flowers in the buttercup is a heritable variation and supposed it to be a unit-character, and so to conform with Mendel's law.

Now, if the tendency to produce double flowers were a simple Mendelian character it could exist in only three conditions, — that of a recessive, that of a homozygous dominant, or that of a heterozygous dominant. But recessives and homozygous dominants are pure, that is, they form only one type of gamete, and selection therefore from among their progeny could produce no new type. As regards the heterozygous dominant type, this would itself be unfixable, and selection could accomplish nothing permanent except by isolating a homozygous type. But such types should all be in evidence within two generations; therefore, if a completely and permanently double type had not been discovered within the five generations covered by the experiment, such a type was not to be expected at all from the material in hand, unless either a wholly new unit-character were introduced or an existing one were profoundly modified. De Vries considers changes of both these sorts possible.

He calls them *mutations,* and regards them as the *sole means* of evolutionary progress. But it is a peculiarity of his mutation theory that it regards only *large* changes in unit-characters as having any permanency, namely, such changes as mean a practical making over of the character. To borrow a figure from Bateson, just as the gas carbon monoxide, CO, may change into a very different gas, — carbon dioxide, CO_2, — by taking up a single atom of oxygen, but can make no less extensive change, since oxygen atoms do not split; so, according to De Vries, a unit-character may not change unless it changes profoundly. Various circumstances may. modify the degree of its expression, but these are without permanent effect, since the character itself remains unchanged.

But there are both *a priori* and experimental grounds for questioning the correctness of De Vries' conclusions. It is known that the chemical compounds within the germ-cells are not so simple in composition as CO and CO_2. They are very complex substances, made up, it is thought, of very many atoms, often hun-

110

dreds in a single molecule. If so, it is quite possible that an atom or two might be transposed in position within the molecule without wholly altering its chemical nature, and that thus slight changes in the germ-plasm might result, which, however, would be as permanent as more profound changes.

The argument of De Vries against any permanent effect of selection in modifying unit-characters has been greatly strengthened by the subsequent work of Johannsen and Jennings. Johannsen has found that if one selects from a handful of ordinary beans the largest seeds and the smallest seeds, and plants these separately, the former will produce beans of larger average size than the latter. Selection here has effect.

But if the selection is made, not from a general field crop of beans, but from those beans borne on one and the same homozygous mother plant, then the progeny of the selected large seed will be no larger than that of the selected small seed. Selection here is without effect.

The different result in the two cases may be explained, according to Johannsen, on

111

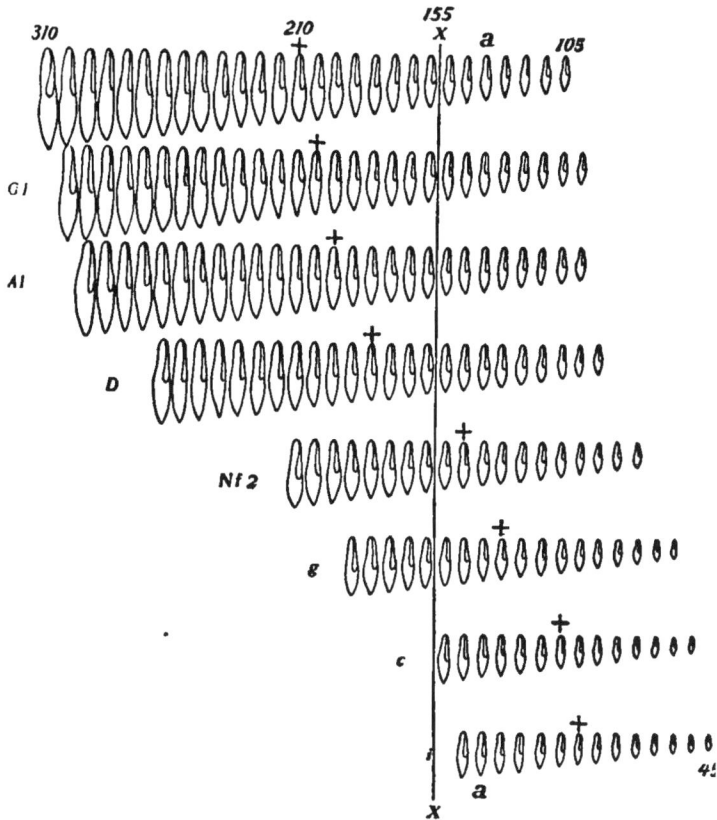

Fig. 40. — Diagram showing the variations in size of eight different races of paramecium. Each horizontal row represents a race derived from single parent individual. The individual showing the mean size in each race is indicated by a cross placed above it. The mean of the entire lot is shown at X — X. The numbers show the measurements in microns. (After Jennings.)

the principle of the "pure line." The progeny of a single self-fertilized homozygous bean plant constitute a pure line. They are all alike, so far as the hereditary transmission of size is concerned, for they are all derived from like gametes. The differences in size which occur among them are due to differences in nutrition, not to germinal differences, and they are not transmitted. But in a mixed population of beans, such as is represented by a field crop, differences of size occur which are due to heredity as well as those which are due to the environment. In the case of the former, selection naturally has effect; in the case of the latter, it does not.

Jennings has obtained similar results in his studies of paramecium, — a one-celled animal which multiplies asexually by dividing into two similar parts. It lives in stagnant water and may be reared in great numbers in a hay-infusion, for it multiplies with great rapidity, dividing two or three times within twenty-four hours. The variations in size which occur in paramecium are shown in Fig. 40.

When from an ordinary culture of parame-

cium Jennings selects the largest and the smallest individuals respectively, he finds that the
descendants of the one lot will be of larger size
than the other. This looks like an effect of selection upon racial size. But if selection is made not
within a mixed population but among the descendants of a single individual, it is found
that the descendants of large individuals are
of no greater average size than those of small
individuals.

The explanation of this fact is to be
found in the existence of what Johannsen
has called pure lines. Jennings has been able
to isolate eight distinct pure lines of paramecium differing in average size, as shown in
Fig. 40. The range of variation in size within
one of these races is great, but if one selects
extremely large or extremely small individuals
within the same pure line, i. e. among the
asexually produced descendants of the same
animal, no change in the average size of the
race is brought about.

A very different result is obtained, however,
if one mixes together several pure lines and then
selects from the mixed race on the basis of size.

The larger animals then produce larger average offspring and *vice versa*. An examination of Fig. 40 will show why. Animals of the same absolute size are there placed in the same vertical row. If, now, one selects from the mixed population only the largest individuals, he will naturally secure representatives of only two or three pure lines, viz. of those lines which are characterized by the largest average size, and which, therefore, will produce large average offspring. If on the other hand he selects extremely small individuals, he will secure representatives of only the smallest races, which naturally will produce small offspring, so that selection seems to be effective in modifying racial size, but in reality it does this by sorting out the elementary constituents of the race.

It is impossible to deny the soundness of the reasoning of Johannsen and Jennings. It is perfectly clear that the effects of selection *should* be more immediate and much greater in the case of a mixed race than in that of a pure line, but is it certain, as assumed by them, that selection is *wholly* without effect in the case of a pure line? We know the effects should be *less*, but are they *nil?* Concerning this matter we are perhaps justified in awaiting further evidence. For in the case of beans and of paramecium alike size is subject to very great variation through the influence of nutrition. Variations due to this cause are naturally not inherited, since the germ-cells are not affected by them, but only the body. But is it not possible that along with the striking size differences due to nutrition there may occur also slight size differences due to germinal variation within the pure line, that is owing to variations in the potency of the same unit-character or combination of unit-characters? To be sure, Johannsen and Jennings have not observed these, but this does not prove their non-existence. Others may yet be able to do so; indeed one case is already on record in which such observations have been made in the case of a small crustacean (or water-flea), Daphnia.

Daphnia is a small transparent animal, about the size of a pin-head, which occurs in enormous numbers in fresh-water lakes and pools,

forming a large part of the food supply of
fresh-water fishes. It multiplies chiefly by the
production of unfertilized eggs, — those which
undergo no reduction and which develop without fertilization into an individual like the
parent. The germinal composition, therefore,
of all descendants produced in this way by the
same mother should be identical, unless germinal composition can be modified in other ways
than by reduction and recombination of unit-
characters. Now the German zoölogist, Woltereck, has shown that, among the offspring
developed from the unfertilized eggs of the
same mother Daphnia, variations do occur
which are heritable, so that if one selects extreme variants he obtains a modified race.
Systematic zoölogists recognize as a generic
distinction between Daphnia and Hyalodaphnia
absence from the latter of the rudimentary eye
found in Daphnia. Woltereck observed that
in a pure line of Hyalodaphnia the rudimentary eye, usually wanting, may occur in
individual cases. He found further that it
occurred in varying degrees of development,
which ranged all the way from a group of

pigmented cells outside the brain, through stages in which cells were present without pigment, and others in which pigment was visible within the brain but no cells outside it were developed, and finally to those in which all traces of the eye had vanished, cells and pigment alike. By selection in three successive generations of the mother having the rudimentary eye best developed offspring were obtained, 90 % of which had the pigmented eye, and which would therefore pass for animals of a wholly different genus. The degree of development of the organ in the last generation was also greater than in the previous generations. Here within a pure line produced by parthenogenesis selection served to augment both the degree of development of an organ and the frequency of its occurrence within the race, a result precisely parallel to that which I obtained some years ago by selection in the case of a rudimentary fourth toe in the guinea-pig. The experiment with Daphnia is not open to the objection that may be offered to the guinea-pig experiment, that it is possibly a result of gametic segregation and recombination, for in Daphnia the reproduction was exclusively by unreduced and unfertilized eggs.

The rudimentary eye of Daphnia is an organ the development of which, so far as observed, is wholly independent of environmental influence; but the case is different with another structure of Daphnia, upon which also Woltereck made observations, namely, a projection or spine borne on the head of the animal. This is not a constant structure, but is sometimes present, sometimes wanting altogether, in the same pure line. In extreme cases it forms a great angular extension of the head forward. To a considerable extent its development is subject to control through the temperature of the surrounding water, but independently of such influence the degree of its development varies and is heritable. Although in general, just as in the experiments of Johannsen and Jennings, selection of animals with the best-developed spine did not increase the degree of development of the organ or the frequency of its occurrence, yet in individual cases such increase was observed, so that the structure occurred in over 50 % of the offspring. In

such cases, then, it would seem that along with
the cases due to environmental influence occurred others due to germinal variation. Although selection of the former would not influence the race permanently, there is every
reason to think that the latter would so influence it, and did in the experiment.

Accordingly the results of Johannsen and
Jennings on the one hand, and of Woltereck
on the other, are not necessarily in opposition
to each other. Woltereck's conclusions agree
with those of Johannsen and Jennings so far
as concerns the great bulk of the variations,
those caused by external influences. All agree
that they are not inherited. Woltereck, however, observes also, what the others have failed
to observe, that along with the non-inherited
variations occur other similar but less numerous ones which are inherited.

My own observations are entirely in harmony with those of Woltereck. Like him, I
find that selection may modify characters. In
several cases I have observed characters at
first feebly manifested gradually improve under
selection until they became established racial

traits. Thus the extra toe of polydactylous
guinea-pigs made its appearance as a poorly
developed fourth toe on the left foot only.
Only 6 % of the offspring of this animal by
normal unrelated mothers were polydactylous,
but among his offspring were some with better
developed fourth toes than the father possessed. Such individuals were selected throughout five successive generations, at the end of
which time a good four-toed race had been
established. It was found in general that those
animals which had best-developed fourth toes
transmitted the character most strongly in
crosses with unrelated normal animals. The
percentage of polydactylous individuals produced in such crosses varied all the way from
0 to 100 %. By selection this percentage was
increased, as was also the degree of development of the fourth toe in crosses.

Another character which made its appearance among our guinea-pigs, at first feebly
expressed, was a silvering of the colored fur,
due to interspersing of white hairs with the
colored ones (see Fig. 37). The first individuals observed to have this character bore

white hairs on the under surface of the body
only. By inbreeding, a homozygous strain of
the silvered animals was soon obtained, one in

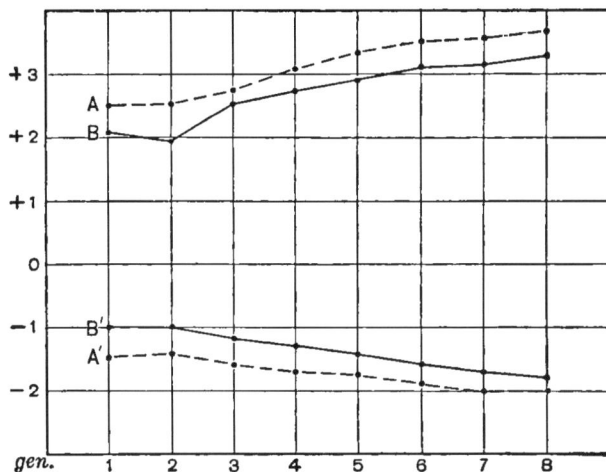

FIG. 41. — Chart showing effects of selection in eight successive
generations upon the color-pattern of hooded rats. *A*,
average condition of the selected parents in the *plus* series;
B, average condition of their offspring. *A*[1], average condition of the selected parents in the *minus* series; *B*[1], average
condition of their offspring.

which all the offspring were silvered to a
greater or less extent. Selection was now
directed toward two ends, — (1) to secure animals which were free from spots of red or
white, a condition which was present in the

original stock, and (2) to secure extensive and uniform silvering on a black background. In both these objects good progress has been made. We have animals which are silvered all over the body except on a part of the head, and the percentage of such well-silvered individuals is relatively high.

But the most extensive selection experiment which I have personally observed is one in which I have been assisted by Dr. John C. Phillips (see Figs. 39 and 41). Selection in this case has been directed toward a modification of the color pattern of hooded rats, — a pattern which is known to behave as a recessive Mendelian character in crosses with either the self (totally pigmented) condition or the so-called Irish (white-bellied) condition found in some other rats. The extreme range of variation among our hooded rats at the outset of this experiment is indicated by the grades — 2 and + 3 of Fig. 39. Selection was now made of the extreme variates in either direction and these were bred separately. Two series of animals were thus established, — one of narrow striped animals, *minus* series; the

other of wide striped, *plus* series. In each
generation the most extreme individuals were
selected as parents; in the narrow series, those
with narrowest stripe; in the wide series, those
with widest stripe.

TABLE I

*Results of Selection for Modification of the Color-pattern
of Hooded Rats.*

	GENERA-TION.	AVERAGE GRADE, PARENTS.	AVERAGE GRADE, OFFSPRING.	NUMBER OF OFF-SPRING.
Plus series.	1	2.50	2.05	150
	2	2.51	1.92	471
	3	2.73	2.51	341
	4	3.09	2.72	444
	5	3.33	2.90	610
	6	3.51	3.09	834
	7	3.53	3.14	874
	8	3.65	3.30	91
				3,815
Minus series.	1	1.46	1.00	55
	2	1.41	1.07	132
	3	1.56	1.18	195
	4	1.69	1.28	329
	5	1.73	1.41	701
	6	1.86	1.56	1252
	7	2.00	1.70	1544
	8	2.03	1.78	713
				4,921

The result of the selection is shown graphically in Fig. 41 (compare Table I). The offspring in the narrow series became with each generation narrower; those in the wide series became with each generation wider, with a single exception. In generation two the wide stock was enlarged by the addition of a new strain of animals. This caused a temporary falling off in the average grade of the young, the two series overlapping for that generation. No new stock was at any other time introduced in either series, the two remaining distinct at all times except in generation two. It will be observed that a change in the average grade of the parents is attended by a corresponding change in the average grade of the offspring. The amount of variability of the offspring is not materially affected by the selection, but the average about which variation occurs is steadily changed, as are also the limits of the range of variation.

The interesting feature of this experiment is the production, as a result of selection, of wholly new grades; in the narrow series, of animals having less pigment than any known type other than the albino; in the wide series, of animals so extensively pigmented that they would readily pass for the '' Irish type,'' which has white on the belly only, but which is known to be in crosses a Mendelian alternative to the hooded type. By selection we have practically obliterated the gap which originally separated these types, though selected animals still give regression toward the respective types from which they came. But this regression grows less with each successive selection and ultimately should vanish, if the story told by these statistics is to be trusted. As yet there is no indication that a limit to the effects of selection has been reached.

From the evidence in hand we conclude that Darwin was right in assigning great importance to selection in evolution; that progress results not merely from sorting out particular combinations of large and striking unit-characters, but also from the selection of slight differences in the potentiality of gametes representing the same unit-character combinations. It is possible to ascribe such differences to little units additional to the recognized larger ones, but if such little units exist, they are indeed very little as well as numerous, and by adding to the effect of the larger ones they produce what amounts to modification of them.

BIBLIOGRAPHY

CASTLE, W. E.
  1906. "The Origin of a Polydactylous Race of Guinea-pigs." *Carnegie Institution of Washington, Publication No. 49*, pp. 17–29.

JENNINGS, H. S.
  1909. "Heredity and Variation in the Simplest Organisms." *The American Naturalist*, 43, pp. 321–337.

JOHANNSEN, W.
  1909. "Elemente der exakten Erblichkeitslehre." G. Fisher, Jena, 516 pp.

DE VRIES, H.
  (See Bibliography to Chapter VI.)

WOLTERECK, R.
  1909. "Weitere experimentelle Untersuchungen über Art-veränderungen, speciell über das Wesen quantitativer Artunterschiede bei Daphniden." *Verh. Deutsch. Zool. Gesellsch.*, pp. 110–172.

CHAPTER VIII

WE shall now discuss a seemingly different type of inheritance from that discovered by Mendel, — one in which the offspring are a true intermediate or *blend* between the parents, and in which the occurrence of segregation has not in all cases been certainly established.

Differences in size between parents have been found to behave in this blending fashion. Rabbits are apparently favorable material in which to study size inheritance, for some races are fully twice as large as others. If a large rabbit is crossed with a small one the young are of intermediate size, and the F_2 offspring show no such segregation into large, small, and intermediate individuals as a simple Mendelian system would demand. For this reason size has been de-

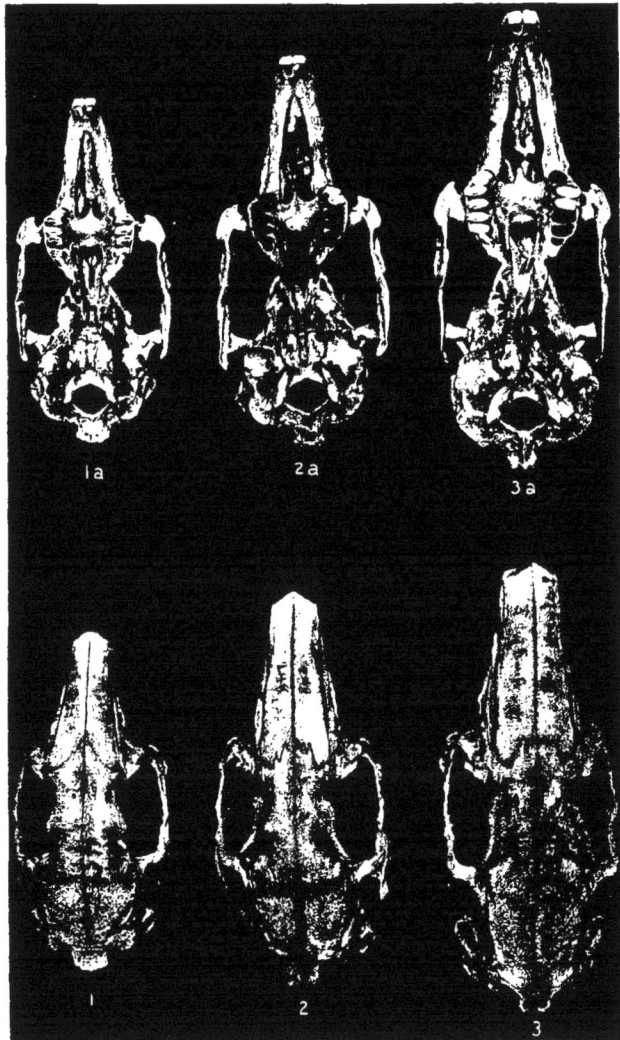

FIG. 42. — Skulls of three rabbits. Father (1 and 1a),
mother (3 and 3a), and son (2 and 2a).

scribed as a non-Mendelian, non-segregating
type of inheritance, but recent discoveries place
this interpretation in doubt. Let us first consider what are the observed facts and after-
ward the interpretation.

Fig. 42 shows the skulls of three rabbits, —
of the father at the left, of the mother at the
right, and that of the son between. Notice the
fully intermediate or blended character of the
son's skull as regards both absolute dimensions
and proportions. The intermediate character
was possessed also by the next generation of
offspring. Now this same cross, while producing a blend in size and ear-length, was
yielding dominance and segregation of coat-
characters. Fig. 43 shows a picture of the
rabbit with the small skull in the cross just
described. He was an albino and his fur was
long. The mother, which had the large skull,
was a sooty-yellow rabbit, with short fur and
long ears (see Fig. 44). The son is shown
in Fig. 45. His fur was black and short, the
albinism and long fur of his father having
become recessive in the cross in accordance
with Mendel's law. The pigmentation is also

intensified in the son; black having been received through the albino parent as a latent factor, which became fully active in the son. The excluded albinism, recessive in the son and his brothers and sisters, all seven of which were similar in character, reappeared among the grandchildren, as, for example, in the one shown in Fig. 46, which was short-haired. Other F_2 offspring were long-haired, some of them being albinos, others being pigmented. But the size and ear-length of the son were intermediate between the sizes and ear-lengths of his parents, and this intermediate character persisted without apparent segregation among the F_2 offspring. The animals in the pictures are unfortunately not all shown on the same scale, but the relative ear-lengths are sufficiently clear.

A Mendelian interpretation of blending inheritance, illustrated in the inheritance of skull-size and ear-length among rabbits, has been suggested by my colleague Dr. East, and by others, an interpretation in which Mendelian dominance is indeed wanting but segregation nevertheless occurs, yet not of a simple kind, involving one or two segregating factors, but involving several such factors. Before entering into this explanation it will be necessary to discuss a further extension of Mendelian principles recently made.

Some modified Mendelian ratios of particular interest have lately been obtained by the Swedish plant-breeder, Nilsson-Ehle (1909, Lunds Universitets Arsskrift) in crossing varieties of wheat of different color. When a variety having brown chaff is crossed with one which has white chaff, the hybrid plants are regularly brown in F_1 and 3 brown : 1 white in F_2, but a particular variety of brown-chaffed wheat gave a different result. In 15 different crosses it gave uniformly a close approximation to the ratio 15 : 1 instead of 3 : 1. The totals are sufficiently large to leave no doubt of this. They are 1410 brown to 94 white, exactly 15 : 1. This is clearly a dihybrid Mendelian ratio, and Nilsson-Ehle interprets it to mean that there exist in this case two independent factors, each of which is able by itself to produce the brown coloration, though no qualitative difference can be detected between them.

A still more remarkable case was observed in crosses between varieties of wheat of different grain-color. Red crossed with white gave ordinarily all red in F_1 and 3 red to 1 white in F_2, but a certain native Swedish sort gave only red (several hundred seeds) in F_2. This result was so surprising that one cross which had yielded 78 grains of wheat in F_2 was followed into F_3, with the following result:

50 plants gave only red seed;   expected 37
 5 "   " approximately 63 R:1 W; "   8
15 "   "   "   15 R:1 W; " 12
 8 "   "   "   3 R:1 W; "   6
 0 "   "   *all* white; "   1

The interpretation given by Nilsson-Ehle is this. The red variety used in this cross bears three independent factors, each of which by itself is able to produce the red character. Their joint action is not different in kind from their action separately, though possibly quantitatively greater. The F_2 generation should contain 1 white seed in 64. It happens that none were obtained in this generation. The next generation should contain in a total of 64 individuals, the sorts actually observed as

FIG. 43

FIG. 44

FIG. 45

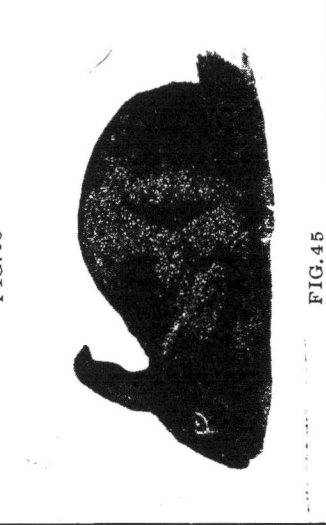

FIG. 46

Fig. 43. — A long-haired, albino rabbit, having erect ears. His skull is shown in Fig. 42 (1 and 1a).

Fig. 44. — A short-haired, sooty yellow rabbit, having lop ears. Her skull is shown in Fig. 42 (3 and 3a).

Fig. 45. — A short-haired, black rabbit, son of the rabbits shown in Figs. 43 and 44. Notice the intermediate ear length. His skull is shown in Fig. 42 (2 and 2a).

Fig. 46. — An F₂ descendant of the rabbits shown in Figs. 44 and 45.

well as a sort which would produce only white seed, the progeny namely of the expected white seed of F_2, but as that was not obtained, the all-white plant of F_3 could not be obtained either. The expected proportions of the several classes in F_3 are given for comparison with those actually obtained. The agreement between expected and observed is so good as to make it seem highly probable that Nilsson-Ehle's explanation is correct. Corroborative evidence in the case of maize has been obtained by Dr. E. M. East (Am. Naturalist, Feb., 1910).

This work introduces us to a new principle which may have important theoretical consequences. If a character ordinarily represented by a single unit in the germ-plasm may become represented by two or more such units identical in character, then we may expect it to dominate more persistently in crosses, fewer recessives being formed in F_2 and subsequent generations. Further, if duplication of a unit tends to increase its intensity, as seems probable, then we have in this process a possible explanation of quantitative variation in characters which are non-Mendelian, or at any rate do not conform with a simple Mendelian system. Consider, for example, the matter of size and skeletal proportions in rabbits. It is perfectly clear from the experiments described that in such cases no dominance occurs, and also that no segregation of a simple Mendelian character takes place, but it is not certain that the observed facts may not be explained by the combined action of several similar but independent factors, the new principle which Nilsson-Ehle has brought to our attention. Let us apply such a hypothesis to the case in hand.

Suppose a cross be made involving ear-lengths of approximately 4 and 8 inches respectively, as in one of the crosses made. The F_1 young are found to have ears about 6 inches long, the mean of the parental conditions, and the F_2 young vary about the same mean condition. If a single Mendelian unit-character made the difference between a 4 inch and an 8 inch ear, the F_2 young should be of three classes as follows:

Classes	4 in.	6 in.	8 in.
Frequencies	1	2	1

(Compare Fig. 47, bottom left.) The grandparental conditions should in this case reappear

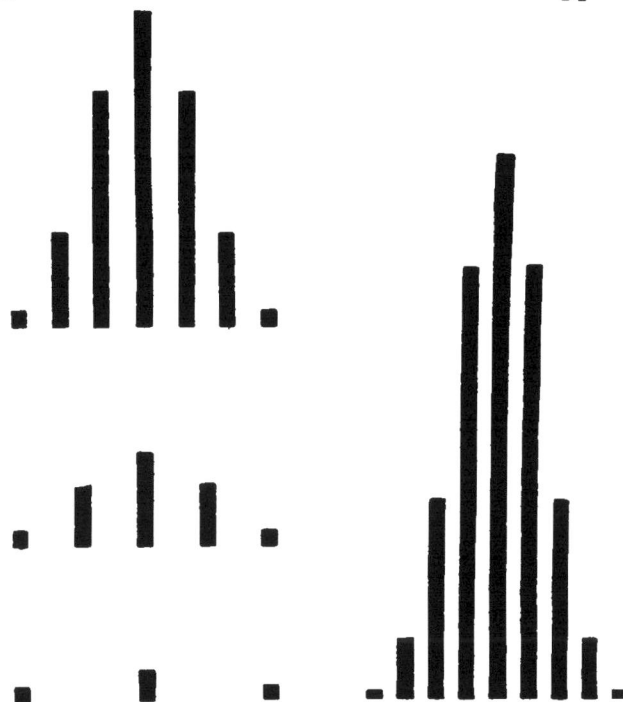

FIG. 47 — Diagrams to show the number and size of the classes
of individuals to be expected from a cross involving Mendelian segregation without dominance. One Mendelian unit
involved, bottom left; two units, middle left; three units,
top left; four units, right.

in half the young. This clearly does not occur in the rabbit experiment. But if two unit-

characters were involved, F_1 would be unchanged, all 6 inches, yet the F_2 classes would
be more numerous, viz., 4, 5, 6, 7, and 8 inches,
and their relative frequencies as shown by the
height of the columns in Fig. 47, middle left, 1, 4,
6, 4, 1. The grandparental states would now reappear in ⅛ of the F_2 young, while ⅜ would be
intermediate. It is certain, however, that in rabbits the grandparental conditions, if they reappear at all, do not reappear with any such
frequency as this.

If three independent size-factors were involved in the cross, the F_1 individuals should all
fall in the same middle group, as before, viz.
6 inches, but the F_2 classes should number
seven, and their relative frequencies would
be as shown in Fig. 47, top left. For 4 independent size-factors, the F_2 classes would be more
numerous still, viz., 9 (Fig. 47, right), and the
extreme ear-size of either grandparent would
be expected to reappear in only one out of 256
offspring, while considerably more than half
of them would fall within the closely intermediate classes included between 5½ and 6½
inches, the three middle classes of the diagram.

With six size-characters, the extreme size of a grandparent would reappear no oftener than once in 4000 times, while with a dozen such independent characters it would recur only once in some 17,000,000 times. It would be remarkable if under such conditions the extreme size were ever recovered from an ordinary cross.

There is one means by which we can determine with certainty whether in a particular case of seemingly blending inheritance segregation does or does not occur, namely, by comparing the variability of the F_1 and the F_2 generations. If segregation does not occur, F_2 should be no more variable than F_1, whereas if segregation does occur, F_2 should be more variable. For, in a segregating system, the F_1 individuals should all fall in a middle, intermediate group, but the F_2 individuals should be distributed also in classes more remote from a strictly intermediate position, that is, they should be more variable. But, in a non-segregating system, F_1 and F_2 individuals alike should fall in the same intermediate group, that is, they should have the same variability.

137

The matter should be easy of determination by observation of considerable numbers of F_1 and F_2 offspring. Investigations are now in progress to test this matter.

My colleague, Dr. East, has found clear evidence that, in maize, size-characters, although they give a blending result in F_1, nevertheless give segregation in F_2. The character to be considered relates to length of ear in corn. A single illustration will suffice. The variation in two pure varieties is shown in the two upper rows of Fig. 48. The " Length " of each class is given in centimetres, its frequency just below at " No. Var.," abbreviation for number of variates. The variation in the F_1 offspring obtained by crossing the two pure varieties is shown in the third row, and that of the F_2 offspring in the lowest row. Note that the variability in the F_1 generation is not increased; its range is intermediate between the range in the parental varieties. In the F_2 generation, however, the variability is so increased that it includes almost the entire range of both parental varieties, together with the intervening region.

In the light of this evidence it is clear that

Length 5 6 7 8
No. Var 4 21 24 8

Length 13 14 15 16 17 18 19 20 21
No. Var 3 11 12 15 26 15 10 7 2

Length 7 8 9 10 11 12 13 14 15 16 17 18 19
No. Var 1 12 12 14 17 9 4

Length 7 8 9 10 11 12 13 14 15 16 17 18 19
No. Var 1 10 19 26 47 73 68 68 39 25 15 9 1

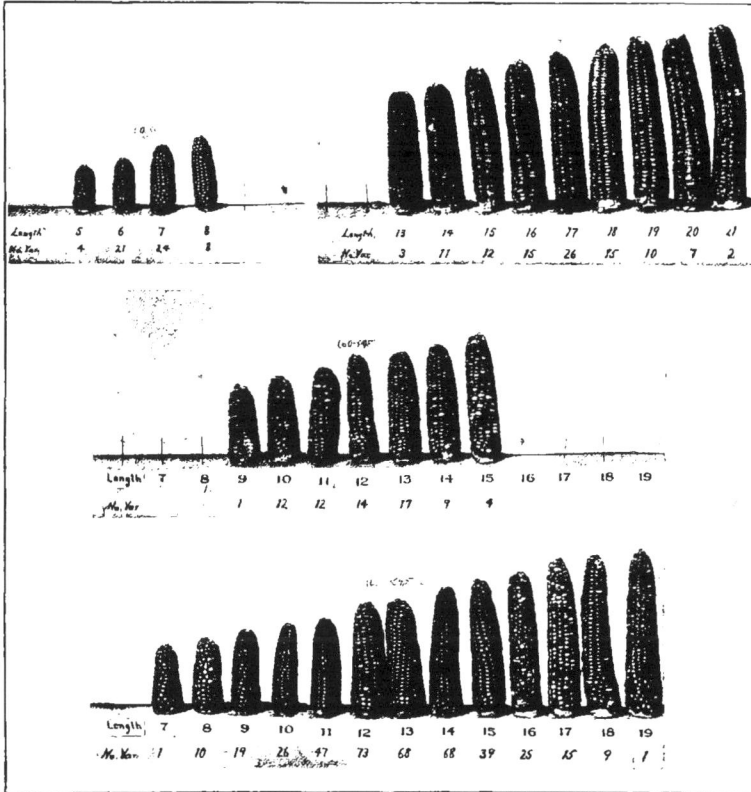

FIG. 48. — Photographs to show variation in ear length of two varieties of maize (upper row), of their F_1 offspring (second row), and of their F_2 offspring (third row).   (After East.)

in maize, seemingly blending is really segregating inheritance, but with entire absence of dominance, and it seems probable that the same will be found to be true among rabbits and other mammals; failure to observe it hitherto is probably due to the fact that the factors concerned are numerous. For the greater the number of factors concerned, the more nearly will the result obtained approximate a complete and permanent blend. As the number of factors approaches infinity, the result will become identical with a permanent blend.

Theoretically it is important to know whether segregating units are involved in inheritance which we call blending; practically it does not matter much, since if these units are only as numerous as six or eight it will be practically impossible to undo the effects of a cross and to recover again the conditions obtaining previous to the cross. The great majority of the offspring both in the first and in subsequent generations following the cross will be strictly intermediate between the conditions crossed whether several units, an infinite number of units, or no units at all are involved.

A practical question of some importance is
how to manipulate simultaneously blending (or
seemingly blending) and Mendelian inheritance.
This must be by a system of line-breeding in
alternate generations, not in successive generations. To test the practicability of this matter
I several years ago set myself the task of com-
bining in one race the large size of some lop-
eared, yellow rabbits which I had, with the
albino character of some small white rabbits
of common race. A first cross produced gray
rabbits of intermediate size, but no white ones.
On inbreeding the gray animals, there were
obtained in F_2 white young of intermediate
size. These were now crossed again with the
original yellow stock, and again colored young
were obtained, but now with $\frac{3}{4}$ of the desired
increase in size. These bred *inter se* again
produced albinos, this time of the $\frac{3}{4}$ size. A
third cross with the original large stock brought
the size up to $\frac{7}{8}$ of that desired, and combined
it in F_2 with the desired albinism. Having
satisfied myself of the correctness of the
method, the experiment was now discontinued.
By further crosses, especially with a fresh lop-

eared stock, to avoid ill-effects of inbreeding, the size could have been still further increased, with judicious selection doubtless up to the extreme size of colored lop-eared rabbits.

The general conclusion to be drawn is that in attempting to combine in one race by cross-breeding characters which exist separately in different races, one should first inquire very carefully how each character, in which the races differ, behaves in transmission, for on the answer to this question should depend the mode of procedure to be chosen.

If simple Mendelian characters only are concerned, nothing is required but to cross the two races and select from the second generation offspring the desired combination. If blending characters only are concerned and F_1 yields the desired blend, this is secure without further procedure, except possibly selection to reduce its variability; but if the desired blend is not yet secured, further back-crossing with one race or the other may be necessary. If, finally, both blending and Mendelian characters are simultaneously involved in a cross, then the method of combined line-breeding and selec-

141

tion in alternate generations, already described, should be adopted.

BIBLIOGRAPHY

CASTLE, W. E.
  1909. (See Bibliography to Chapter V.)

EAST, E. M.
  1910. "A Mendelian Interpretation of Variation that is Apparently Continuous." *American Naturalist*, 44, pp. 65–82.

EMERSON, R. A.
  1910. "The Inheritance of Sizes and Shapes in Plants." *American Naturalist*, 44, pp. 739–746.

NILSSON-EHLE, H.
  1909. "Kreuzungsuntersuchungen an Hafer und Weizen." *Lunds Universitets Arsskrift*, 5, No. 2, 122 pp.

CHAPTER IX

W HAT is the probable source of the evil effects which have been frequently observed to follow inbreeding?

By inbreeding we mean the mating of closely related individuals. As there are different degrees of relationship between individuals, so there are different degrees of inbreeding. The closest possible inbreeding occurs among plants in what we call self-pollination, in which the egg-cells of the plant are fertilized by pollen-cells produced by the same individual. A similar phenomenon occurs among some of the lower animals, notably among parasites. But in all the higher animals, including the domesticated ones, such a thing is impossible because of the separateness of the sexes. For here no individual produces *both* eggs and sperm. The

143

nearest possible approach to self-pollination is in such cases the mating of brother with sister, or of parent with child. But this is less close inbreeding than occurs in self-pollination, for the individuals mated are not in this case *identical* zygotes, though they may be *similar* ones.

It has long been known that in many plants self-pollination is habitual and is attended by no recognizable ill-effects. This fortunate circumstance allowed Mendel to make his remarkable discovery by studies of garden-peas, in which the flower is regularly self-fertilized, and never opens at all unless made to do so by some outside agency. Self-pollination is also the rule in wheat, oats, and the majority of the other cereal crops, the most important economically of cultivated plants. Crossing can in such plants be brought about only by a difficult technical process, so habitual is self-pollination. And crossing, too, in such plants is of no particular benefit, unless by it one desires to secure new combinations of unit-characters.

In maize, or Indian corn, however, among the cereals, the case is quite different. Here enforced self-pollination results in small unproductive plants, lacking in vigor. But racial vigor is fully restored by a cross between two depauperate unproductive individuals obtained by self-fertilization, as has recently been shown by Shull. This result is entirely in harmony with those obtained by Darwin, who showed by long-continued and elaborate experiments that while some plants do not habitually cross and are not even benefited by crossing, yet in many other plants crossing results in more vigorous and more productive offspring; that further, the advantage of crossing in such cases has resulted in the evolution in many plants of floral structures, which insure crossing through the agency of insects or of the wind.

In animals the facts as regards close fertilization are similar to those just described for plants. Some animals seem to be indifferent to close breeding, others will not tolerate it. Some hermaphroditic animals (those which produce both eggs and sperm) are regularly self-fertilized. Such is the case, for example, with many parasitic flat-worms. In other cases self-fertilization is disadvantageous. One such case I was able to point out some fifteen years

145

ago, in the case of a sea-squirt or tunicate,
Ciona. The same individual of Ciona produces
and discharges simultaneously both eggs and
sperm, yet the eggs are rarely self-fertilized,
for if self-fertilization is enforced by isolation
of an individual, or if self-fertilization is
brought about artificially by removing the eggs
and sperm from the body of the parent and
mixing them in sea-water, very few of the
eggs develop, — less than 10 %. But if the
eggs of one individual be mingled with the
sperm of any other individual whatever, practically all of the eggs are fertilized and
develop.

In the great majority of animals, as in many
plants, self-fertilization is rendered wholly impossible by separation of the sexes. The same
individual does not produce *both* eggs and
sperm, but only one sort of sexual product.
But among sexually separate animals the same
degree of inbreeding varies in its effects. The
closest degree, mating of brother with sister,
has in some cases no observable ill-effects.
Thus, in the case of a small fly, Drosophila,
my pupils and I bred brother with sister for

fifty-nine generations in succession without obtaining a diminution in either the vigor or the fecundity of the race, which could with certainty be attributed to that cause. A slight diminution was observed in some cases, but this was wholly obviated when parents were chosen from the more vigorous broods in each generation. Nevertheless crossing of two inbred strains of Drosophila, both of which were doing well under inbreeding, produced offspring superior in productiveness to either inbred strain. Even in this case, therefore, though inbreeding is tolerated, cross-breeding has advantages.

In the case of many domesticated animals, it is the opinion of experienced breeders, supported by such scientific observations as we possess, that decidedly bad effects follow continuous inbreeding. Bos ('94) practiced continuous inbreeding with a family of rats for six years. No ill-effects were observed during the first half of the experiment, but after that a rapid decline occurred in the vigor and fertility of the race. The average-sized litter in the first half of the experiment was about 7.5, but in the last year of the experiment it had fallen to 3.2, and many pairs were found to be completely sterile. Diminution in size also attended the inbreeding, at the end amounting in the case of males to between 8 and 20 %.

Experiments made by Weismann confirm those of Bos as regards the falling off in fertility due to inbreeding. For eight years Weismann bred a colony of mice started from nine individuals, — six females and three males. The experiment covered 29 generations. In the first 10 generations the average number of young to a litter was 6.1; in the next 10 generations, it was 5.6; and in the last 9 generations, it had fallen to 4.2. But sweeping generalizations cannot be drawn from these cases. Each species of animal must probably be tested for itself before we shall know what the exact effects of inbreeding are in that case. In guinea-pigs, a polydactylous race built up by the closest inbreeding out of individuals all descended from one and the same individual has now been in existence for ten years. It consists of one of the largest and most vigorous strains of guinea-pigs that I have ever

148

seen, and has shown no indications of diminished fertility.

In the production of pure breeds of sheep, cattle, hogs, and horses inbreeding has frequently been practiced extensively, and where in such cases selection has been made of the more vigorous offspring as parents, it is doubtful whether any diminution in size, vigor, or fertility has resulted. Nevertheless it very frequently happens that when two pure breeds are *crossed*, the offspring surpass either pure race in size and vigor. This is the reason for much cross-breeding in economic practice, the object of which is not the production of a new breed, but the production for the market of an animal maturing quickly or of superior size and vigor. The inbreeding practiced in forming a pure breed has not of necessity *diminished* vigor, but a cross does temporarily (that is in the F_1 generation) *increase* vigor above the normal. Now why should inbreeding unattended by selection decrease vigor, and cross-breeding increase it? We know that inbreeding tends to the production of homozygous conditions, whereas cross-breeding tends to

11   149

produce heterozygous conditions. Under self-pollination for 1 generation following a cross, *half* the offspring become homozygous; after 2 generations, ¾ of the offspring are homozygous; after 3 generations ⅞ are homozygous, and so on. So if the closest inbreeding is practiced there is a speedy return to homozygous, pure racial conditions. We know further that in some cases at least heterozygotes are more vigorous than homozygotes. The heterozygous yellow mouse is a vigorous lively animal; the homozygous yellow mouse is so feeble that it perishes as soon as produced, never attaining maturity. Cross-breeding has, then, the same advantage over close-breeding that fertilization has over parthenogenesis. It brings together differentiated gametes, which, reacting on each other, produce greater metabolic activity.

Inbreeding, also, by its tendency to secure homozygous combinations, tends to bring to the surface latent or hidden recessive characters. If these are in nature defects or weaknesses of the organism, such as albinism and feeble-mindedness in man, then inbreeding is distinctly bad. Existing legislation against the marriage of near-of-kin is, therefore, on the whole, biologically justified. On the other hand, continual crossing only tends to *hide* inherent defects, not to exterminate them; and inbreeding only tends to bring them to the surface, not to *create* them. We may not, therefore, lightly ascribe to inbreeding or intermarriage the *creation* of bad racial traits, but only their manifestation. Further, any racial stock which maintains a high standard of excellence under inbreeding is certainly one of great vigor, and free from inherent defects.

The animal breeder is therefore amply justified in doing what human society at present is probably not warranted in doing, — viz. in practicing close inbreeding in building up families of superior excellence and then keeping these pure, while using them in crosses with other stocks. For an animal of such a superior race should have only vigorous, strong offspring if mated with a healthy individual of any family whatever, within the same species. For this reason the production of " thoroughbred " animals and their use in

151

crosses is both scientifically correct and com-
mercially remunerative.

BIBLIOGRAPHY

Bos, Ritzema.
    1894. "Untersuchungen ueber die Folgen der Zucht in
        engster Blutverwandtschaft." *Biol. Centrbl.*, 14, pp. 75–
        81.

Castle, W. E., Carpenter, F. W., Clark, A. H., Mast,
    S. O., and Barrows, W. M.
    "The Effects of Inbreeding, Cross-breeding and Selec-
        tion upon the Fertility and Variability of *Drosophila.*"
        Proc. Amer. Acad. Arts and Sci., 41, pp. 731–786.

Guaita, G. von.
    1898. "Versuche mit Kreuzungen von verschiedenen
        Rassen des Hausmaus." *Ber. naturf. Gesellsch. zu
        Freiburg*, 10, pp. 317–332. [Contains observations of
        Weismann.]

CHAPTER X

THE value of a domesticated animal often depends in considerable measure on its sex. Therefore, if a means could be devised for controlling the sex of offspring, it would be of great economic value to the breeder. Endless attempts have been made to do this, and occasionally a claim of success has been made, but none of these claims has withstood the test of critical analysis or experiment. The hypotheses advanced to explain how sex may be controlled have been of the most varied character. In some the determination of sex has been supposed to inhere in the nature of the parents, in others it is referred to conditions of the gametes themselves.

Relative age or vigor of the parents have been supposed to influence sex in various ways. The same idea has been advanced regarding

the gametes themselves, it being supposed that
early or late fertilization of the egg might
influence its sex. Experimental evidence, however, as to these several hypotheses is wholly
negative, when one eliminates other possible
factors from the experiment. Everything
points to the conclusion that sex rests in the
last analysis upon gametic differentiation, just
as the color of a guinea-pig in a mixed race
of blacks and whites depends upon whether the
gametes which unite to produce it carry black
or white. As the heterozygous black guinea-
pig forms black-producing and white-producing
gametes in equal numbers, so there is reason
to think male-producing and female-producing
gametes are formed in equal numbers by the
parent, in many cases at least. But is it not
possible that there may exist individuals which
produce the two sorts of gametes in unequal
numbers, and so would have a tendency to
produce more offspring of one sex than of the
other? Perhaps so, though we have no evidence that such a condition, if it does exist, is
transmitted from one generation to another.
On this point I made experimental observa-

✓ tions upon guinea-pigs extending over a series of years. Oftentimes I found an individual that produced more offspring of one sex than of the other, but this was probably due merely to chance deviations from equality. I could get no evidence that the condition was inherited, though the experiment was continued through as many as seven generations, including several hundred offspring.

The essential difference between a female and a male individual is that one produces eggs, the other sperm. All other differences are secondary and dependent largely upon the differences mentioned. If in the higher animals (birds and mammals) the sex-glands (*i. e.* the egg-producing and sperm-producing tissues) are removed from the body, the superficial differences between the sexes largely disappear. In insects, however, the secondary sex-characters seem to be for the most part uninfluenced by presence or absence of the sex-glands. Their differentiation occurs independently though simultaneously with that of the sex-glands.

The egg or larger gamete (the so-called

macro-gamete) in all animals is non-motile and contains a relatively large amount of reserve-food material for the maintenance of the developing embryo. This reserve-food material it is the function of the mother to supply. In the case of some animals, for example flatworms and mollusks, the food-supply of the embryo is not stored in the egg-cell itself, but in other cells associated with it, and which break down and supply nourishment to the developing embryo derived from the fertilized egg. Again, as in the mammals, the embryo may derive its nourishment largely from the maternal tissues, the embryo remaining like a parasite within the maternal body during its growth, feeding by absorption. But in all cases alike the mother supplies the larger gamete and the food-material necessary to carry the zygote through its embryonic stages. The father, on the other hand, furnishes the bare hereditary equipment of a gamete, with the motor apparatus necessary to bring it into contact with the egg-cell, but without food for the developing embryo produced by fertilization. The gamete furnished by the father is

156

therefore the smaller gamete, the so-called *micro-gamete*.

From the standpoint of metabolism, the female is the more advanced condition; the female performs the larger function, doing all that the male does in furnishing the material basis of heredity (a gamete), and in addition supplying food for the embryo. As regards the reproductive function, the female is the equivalent of the male organism, plus an additional function, — that of supplying the embryo with food. When we come to consider the structural basis of sex, we find reasons for thinking that here, too, the female individual is the equivalent of the male plus an additional element. The conclusion has very naturally been drawn that if a means could be devised for increasing the nourishment of the egg or embryo, its development into a female should be thereby insured, while the reverse treatment should lead to the production of a male. But in practice this *a priori* expectation is not fulfilled. Better nourishment of the mother may lead to the production of *more eggs*, but not of more *female offspring*, as has

157

been repeatedly demonstrated by experiment. Also poor nutrition of the mother may diminish the number of eggs which she liberates, but will not increase the proportion of males among the offspring produced.

An excellent summary of evidence on this point was made by Cuénot in 1900. Attempts to influence the sex of an embryo or larva by altered nutrition of the embryo or larva itself have proved equally futile. Practically the only experimental evidence of value in favor of this idea has been derived from the study of insects, and this is capable of explanation on quite different grounds from those which first suggest themselves. It has sometimes been observed, as by Mary Treat for example, that a lot of insects poorly fed produce an excess of males. In such lots, however, the mortality is commonly high, and more females die than males, because the female is usually larger and requires more food to complete its development. The fallacy in concluding from such evidence that scanty nutrition causes individuals which would otherwise become females to develop into males was indicated years ago by Riley.

Nevertheless an argument for the artificial control of sex based on such evidence is from time to time brought forward, as, for example, a few years since by Schenk. The latest advocate of sex-control by artificial means is an Italian, Russo (1909). He claims in the case of rabbits that by feeding the mother on lecithin or by injections of lecithin, the proportion of female births may be increased. His evidence in support of this claim is, however, wholly inadequate, and two independent repetitions of his experiments, made by Basile in Italy and by Punnett in England, have given entirely negative results.

An alternative hypothesis concerning the determination of sex has been steadily gaining ground during the last ten years, that sex has its beginning in gametic differentiation and is finally determined beyond recall in the fertilized egg by the nature of the uniting gametes. Instructive in this connection is a study of parthenogenesis, — reproduction by unfertilized eggs. But before entering upon this, it may be well to review briefly the changes which regularly take place in the egg which is to be

159

fertilized, and compare with this the changes which occur in eggs not to be fertilized.

In each cell of the ordinary animal there occurs a characteristic number of bodies called chromosomes. We do not know that they are any more important than other cell constituents, but we know their history better. These are contained in the nucleus of the cell, and at the time of nuclear division they are found at the equator of the division spindle. For example, in the egg of the mouse (Fig. 4, *A*), the nucleus is seen to be in the spindle stage, and its chromosomes are gathered together at the equator of the spindle. There each of them regularly splits in two, and one derivative goes to either end of the spindle, and so into one of the daughter-nuclei. Thus each new nucleus has, as a rule, the same chromosome composition as the nucleus from which it was derived.

But the egg which is to be fertilized undergoes two nuclear divisions in succession, in only one of which do the chromosomes split (see Fig. 4, *A–D*). In the other division the chromosomes separate into two groups without splitting, and each group goes into a different cell product. Consequently, in each of these products the number of chromosomes is reduced to half what it is in the cells of the parental body. Thus in the egg of the mouse, by maturation, the number of chromosomes becomes reduced from about twenty-four to about twelve.

Similar changes occur in the developing sperm-cell (see Fig. 5). Starting with the double or 2 N chromosome number, there are formed by two nuclear divisions, with only one splitting of chromosomes, four cells, each with the reduced or simplex number of chromosomes, N. Consequently, when the sperm enters the egg at fertilization it brings in a group of N chromosomes (in the mouse apparently twelve), which, added to the egg-contribution of N chromosomes, brings the number in the new organism again up to 2 N (in the mouse twenty-four).

Now, as regards the maturation of parthenogenetic eggs, those which are to develop without having been fertilized, three categories of cases deserve separate discussion. The simplest of these in many respects is found among the social hymenoptera (ants, bees, and wasps).

See Fig. 49, left column. The eggs are, so far
as we can discover, all of a single type. They

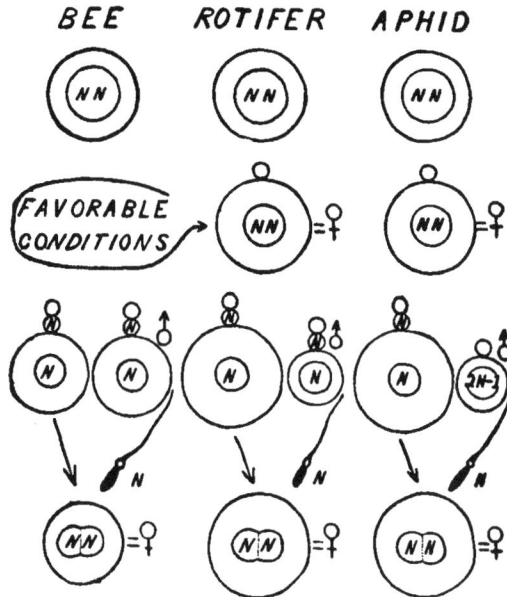

FIG. 49. — Diagram of sex-determination in parthenogenesis.
First row, nuclear condition of the parthenogenetic mother;
second row, of her eggs when they develop without reduction,
after forming a single polar cell; third row, condition of the
eggs after complete maturation — the fertilized egg in each
case produces a male; fourth row, nuclear condition of the
fertilized egg, always a female.

undergo maturation in the manner already described, the chromosomes being reduced to the

N or simplex number. The eggs of most animals, after they have undergone reduction, are incapable of development unless fertilized, but those of the hymenoptera may develop either fertilized or unfertilized. In the former case a female is produced, in the latter a male. The simplex, or N condition is in this case the male, the duplex or 2 N condition is the female, naturally the one of higher metabolic activity, the one which forms the macro-gametes.

In an earlier chapter I explained how the development of the sperm-cells in a male having the reduced or simplex number of chromosomes differs from that in the ordinary male. Reference to Fig. 8 may help to recall this. The cells of the male are in this case already in the reduced or simplex condition, N. In the production of the sperms the reducing division is omitted so far as nuclear components are concerned, so that each sperm formed contains the full simplex chromosome number, N. If it were less, the gamete formed would perhaps not be capable of transmitting all the hereditary characteristics of an individual.

A second category of cases (Fig. 49, middle

column) is represented by such simple aquatic organisms as rotifers and small crustacea, like Daphnia. In these parthenogenesis occurs exclusively, when the food supply is very abundant and conditions otherwise favorable, whereas reproduction by fertilized eggs occurs only when external conditions, including food-supply, are not good. Under favorable conditions only female offspring are produced. The conclusion has naturally but erroneously been drawn that good nutrition in itself favors the production of females in animals generally, which is not true. The egg produced by Daphnia, or by a rotifer, under optimum conditions *does not undergo reduction* (see Fig. 49, second row). It remains in the 2 N condition, forming but a single polar cell. It is therefore unprepared for fertilization, and in fact it is not fertilized. Its sex is like that of the animal which formed it, female. Under unfavorable conditions, however, the eggs of the rotifer and of Daphnia do not begin development until they have undergone maturation. They are also of two sizes (Fig. 49, third row), — small eggs, which develop without fertilization and which form

males, and large eggs, which require fertilization, and which form females. In this
category of cases, as in that of the hymenoptera, the egg which develops in the 2 N condition, either from failure of reduction to occur
in maturation or from fertilization following
reduction, forms a female; whereas the egg
which develops in the N condition forms a
male.

In a third category of cases there is a quantitative difference in chromatin between male
and female, just as in the foregoing cases, but
this does not amount to a whole set of chromosomes, N, but to only a partial set, one or two
chromosomes (see Fig. 49, right column). This
category of cases occurs in plant-lice (aphids
and phylloxerans); evidence of its existence
rests chiefly on recent observations made by
von Baehr and Morgan. Females are formed
by parthenogenesis without reduction, occurring
under favorable conditions, just as in the case
of rotifers. Females are also formed by fertilization following reduction under unfavorable conditions, just as in rotifers. In both
cases the female is 2 N. Males arise only by

parthenogenesis under unfavorable conditions, just as in rotifers, but the reduction which occurs before development begins is partial only. A whole set, N, of chromosomes is not eliminated in maturation, but only 1 or 2 chromosomes. Hence the male condition here is $2N - 1$ or $- 2$. The condition of the gametes formed, however, is N in both sexes. In spermatogenesis, division of the germ-cells takes place into N and $N - 1$ daughter cells, but the latter degenerate (like the non-nucleated cells of the bee and wasp), and only the former produce spermatozoa. Hence in fertilization only $2N$ zygotes are produced, which are invariably female.

Summarizing the three categories described, we may say that in all known cases of parthenogenesis, the female is in the duplex, $2N$ condition, the male in the simplex (N) or partially duplex condition ($2N - 1$, or $2N - 2$). *The female in all cases has the greater chromatin content.*

In a great many insects and other arthropods, which are not parthenogenetic, it is known that, although the male, like the female,

166

develops only from a fertilized egg, nevertheless the male possesses fewer chromosomes than

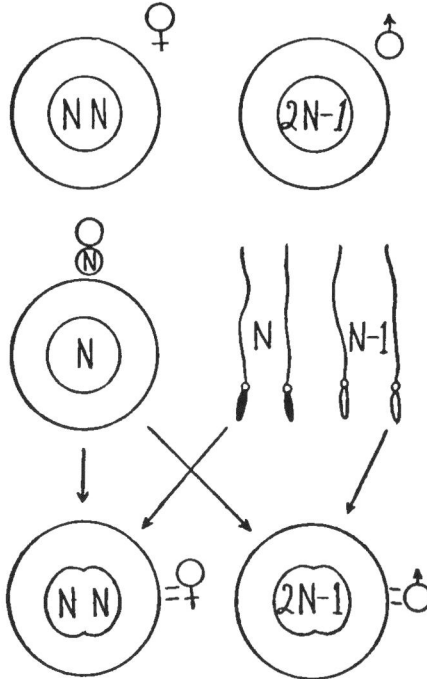

FIG. 50. — Diagram of sex-determination when the female is
homozygous, the male heterozygous.

the female. In such cases the female forms,
as in cases of parthenogenesis, only N gametes,
but the male forms gametes of two sorts, N and

N — 1 or N — 2 (see Fig. 50). In consequence zygotes of two sorts result, — those which are 2 N, female, and those which are 2 N — 1 or 2 N — 2, male. Thus in the squash-bug, Anasa-tristis, according to Wilson, the mature egg contains 11 chromosomes, the spermatozoa either 10 or 11 chromosomes, the two sorts being equally numerous.

Egg 11 + sperm 11 produces a zygote 22 (2N), a female;
Egg 11 +   "   10   "   "   "   21 (2N–1), a male.

N in this species = 11; 2 N = 22, the female; 2 N — 1 = 21, the male. Males and females are therefore approximately equal in number, as in most animals where the two sexes are not sub-ject to unequal mortality. In the Mendelian sense the female is in such cases a homozygote, the male a heterozygote. The sex of an indi-vidual in such cases depends upon which sort of a sperm chances to enter the egg.

But the experimental evidence indicates that both as regards sex and as regards heritable characters correlated with sex, these relations may in some cases be reversed, the female being heterozygous, the male homozygous. In such cases there is reason to think that structurally the male is 2 N but the female 2 N +. That is, the female is still the equivalent of the male *plus* some additional element and function. A structural basis in the chromosomes for such a condition has been described by Baltzer in the case of the sea-urchin. He found the regular duplex number of chromosomes in the male; but in the female, while the number was the same, one of the chromosomes was larger than its mate, having an extra or odd element attached to it. In such a case the gametes formed by the male would all be N, but those formed by the female would be of two sorts equally numerous, viz. N and N + (see Fig. 51). Egg N fertilized by sperm N would produce a zygote 2 N, a male; egg N + fertilized by sperm N would produce a zygote 2 N +, a female. Hence here, as in other animals, the sexes would be approximately equal, but the sex of a particular individual would depend upon which sort of egg gave rise to it.

Upon the existence, as in the foregoing cases, of an unpaired or odd structural element in the egg, may perhaps depend the explanation of a curious sort of heredity known as sex-limited heredity.

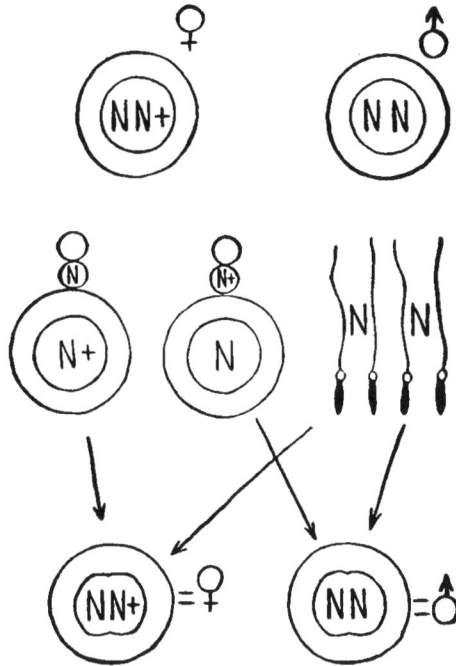

FIG. 51. — Diagram of sex-determination when the female is heterozygous, the male homozygous.

Every one who knows anything about poultry is acquainted with the popular American breed known as the barred Plymouth Rock. In this breed the feathers are marked with alternate bars of darker and lighter black. Pure barred Rocks breed true, but when crossed with other breeds, the male proves to be homozygous, the female, heterozygous in barring. For the male Rock crossed with a non-barred breed produces only barred offspring in both sexes, but the female Rock crossed with the same non-barred breed produces offspring approximately half of which are barred, the other half being non-barred. Further, the barred individuals in this cross are invariably males, the non-barred ones being females. Accordingly, the distribution of barring and non-barring in the cross is sex-limited.

The barred offspring produced by a cross between barred Plymouth Rocks and a non-barred breed, whether those offspring are males or females, prove to be heterozygous in barring, as we should expect, the barring factor having been received only from one parent, the barred one. Further, the *non-barred* offspring produced by a barred Rock female crossed with a non-barred breed, do not transmit barring, hence they are pure recessives as regards barring. Hence, also, we are forced to conclude, as already suggested, the female of the pure barred Rock breed is heterozygous as regards barring, and transmits the character only to her male offspring, her female offspring (if the father is non-barred) neither being themselves barred nor being able to transmit barring.

A pure Plymouth Rock race breeds true to barring merely because all its *males* are pure, for the females are not pure. This is shown by the following experiment. If a heterozygous barred male, produced by a cross between a Rock and a non-barred breed, is crossed with barred females, either those of a pure Rock race or those produced by a cross, the result is the same. The male offspring are all barred; the females, half of them barred, half non-barred. This result shows that all barred females alike are heterozygous in barring.

Sex-limited inheritance such as this finds at the present time its most probable explanation in the existence in the egg of an extra or *plus* element never found in the sperm, this element pairing with the sex-limited character in the reduction division. Thus, in the barred Rock, calling barring B, the male of pure race is plainly B B and every sperm is B. But the female clearly contains only one B and can-

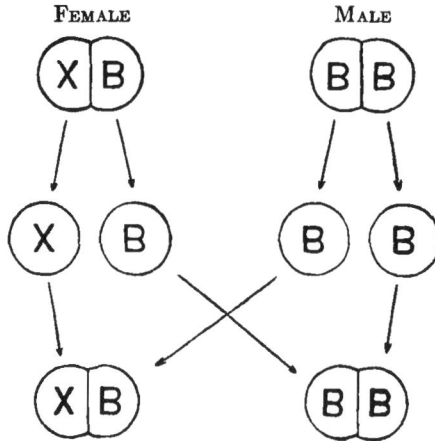

FIG. 52. — Diagram of sex-limited inheritance when the female is a heterozygote, as in barred fowls. X, female sex determiner; B, barring.

not be made to contain two. Perhaps a second B is kept out by some structural element, X, the distinctive structural element of the female individual. Then the eggs will be of two sorts: B and X. Since the sperms are all B, the first type of egg when fertilized will contain B B, a homozygous barred individual and a male, since it lacks X; the second type will contain B X, a bird heterozygous in barring, and a female, since it contains X. This agrees with the experimental result (see Fig. 52).

A heterozygous barred male will form two kinds of sperm, only one of which will contain B. If such a male be mated with a barred female, four sorts of zygotes should result, as follows:

Gametes of heterozygous barred male = B and −,
Gametes of barred female = B and X,
Zygotes = B·B (homozygous barred male); B·− (heterozygous barred male), B·X (barred female), and −·X (non-barred female).

The observed result of this cross accords fully with the foregoing expectation.

The sex-limited inheritance of barring in fowls may be explained, as we have just seen, on the assumption that the female is the heterozygous sex. The same is true of sex-limited inheritance in canary-birds and in the moth, *Abraxas*, according to Bateson and Doncaster. But these relations are exactly reversed in the pomace-fly, *Drosophila ampelophila*, according to Morgan.

In Drosophila the female is apparently homozygous as regards some cell-structure, X, which in the male is never represented more than once. Accordingly the formula of

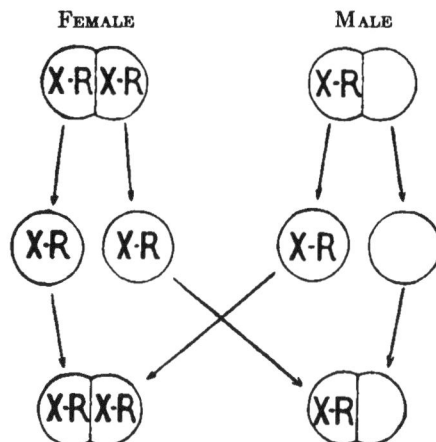

FIG. 53. — Diagram of sex-limited inheritance when the female is a homozygote, as in the red-eyed Drosophila. X, sex-determiner; R, red-eyes.

the female is in such cases X X; that of the male, X —. Now the sex-limited characters in Drosophila seem to be bound up with the X structure, not repelled by it, as is barring in fowls. Accordingly, a sex-limited character may be represented *twice* in the female Drosophila, but only *once* in the male; or in other

175

words, the female may be homozygous as regards a sex-limited character, but the male can
only be heterozygous (see Fig. 53).

Drosophila normally has red eyes, but the
redness of the eye is a distinct unit-character,
sex-limited in heredity. Further males are
regularly heterozygous in this character, while
females are homozygous. For Morgan has obtained a race in which the eyes are white,
owing to the loss of the red character; and
reciprocal crosses of this race with ordinary
red-eyed animals yield different results. The
red-eyed female crossed with a white-eyed male
produces only red-eyed offspring, but the red-
eyed male crossed with a white-eyed female
produces offspring only half of which are red-
eyed, viz. the females, whereas the males are
white-eyed.

These different results in the two cases apparently come about as follows:

First case.

Gametes of red-eyed female = X-R and X-R,
Gametes of white-eyed male = X and —,
Zygotes = X·X-R (red-eyed female), and = —·X-R
(red-eyed male).

Second case.

Gametes of white-eyed female = X and X,
Gametes of red-eyed male = X-R and −,
Zygotes = X·X-R (red-eyed female), and −·X
(white-eyed male).

A short condition of the wings in Drosophila,
which renders the animal incapable of flight,
is likewise sex-limited in heredity, as has been
shown by Morgan. By crossing two races of
Drosophila, each of which possessed a different
sex-limited character, Morgan has been able to
combine the two characters in a single race.
Thus was obtained a race both white-eyed and
short-winged. The synthesis cannot be made
originally in a male individual, but only in a
female. For only in the female can the two
characters be brought together, each associated
with a different X, since in the male only one
X is present. Although each sex-limited character seems to be attached to or bound up with
an X structure, it evidently has a material basis
distinct from X. Otherwise it would not be
possible for the character to leave one X and
attach itself to the other, as apparently takes
place in the female when the combination of

two sex-limited characters in the same gamete is secured through a cross. The combination is apparently secured in this way:

Gametes uniting, X-R and X–L,
Zygote formed, X-R·X-L,
Its gametes, X-R and X-L, or X-R-L and X.

One of the uniting gametes, X-R, is formed by the red-eyed, short-winged parent; the other, X-L, is formed by the long-winged, white-eyed parent. The zygote resulting is a red-eyed individual, since it contains R; it is long-winged, since it contains L; it is a female, since it contains two Xs. Now, its gametes are of four sorts, as indicated. The first two sorts result from simple separation of the two Xs, each with its associated character, R in one case, L in the other. But the third sort could result only from the attachment of R and L to the same X, leaving the other X without either R or L as the fourth kind of gamete. This kind, which transmits neither red eyes nor long wings, would represent the new gametic combination,—white-eyed and with short wings.

The experimental evidence that gametes of

these four sorts are formed by females of the
origin described is as follows: — When such a
female is mated with a long-winged, white-eyed
male, there are obtained female offspring, all
of which are long-winged, but half of them are
red-eyed, half white-eyed. The male offspring,
however, are of four sorts, viz. red short, white
long, red long, and white short. This result
harmonizes with the hypothesis advanced. For
if the gametes of the female are X-R, X-L,
X-R-L, and X, and those of the male are X-L
and —, then the following combinations should
result:

X-L· X-R, red long female,
X-L· X-L, white long female
X-L· X-R-L, red long female,
X-L· X   , white long female,

——· X-R, red short male,
——· X-L, white long male,
——· X-R-L, red long male,
——· X   , white short male.

This expected result accords with that actually obtained by Morgan.

Color-blindness in man is a sex-limited character, the inheritance of which resembles that

of white eyes or short wings in Drosophila, rather than of barring in poultry.

Color-blindness is much commoner in men than in women. A color-blind man, however, does not transmit color-blindness to his sons, but only to his daughters, the daughters, however, are themselves normal provided the mother was; yet they transmit color-blindness to half their sons. A color-blind daughter could be produced, apparently, only by the marriage of a color-blind man with a woman who transmitted color-blindness, since the daughter to be color-blind must have received the character from *both* parents, whereas the color-blind son receives the character only from his mother.

Color-blindness is apparently due to a defect in the germ-cell, — absence of something normally associated there with an X-structure, which is represented twice in woman, once in man. Color-blindness follows, therefore, in transmission the scheme shown in Fig 53.

If, as has been suggested, the determination of sex in general depends upon the inheritance of a Mendelian factor differentiating the sexes,

'it is highly improbable that the breeder will
ever be able to control sex. Male and female
zygotes should forever continue to be produced
in approximate equality, and consistent inequal-
ity of male and female births could result only
from greater mortality on the part of one sort
of zygote than of the other. Only in parthenogenesis can man at will control sex, and until
he can produce artificial parthenogenesis in the
higher animals, he can scarcely hope to control sex in such animals.

Negative as are the results of our study of sex-
control, they are perhaps not wholly without
practical value. It is something to know our
limitations. We may thus save time from
useless attempts at controlling what is uncon-
trollable and devote it to more profitable em-
ployments.

BIBLIOGRAPHY

Bateson, W.
   1909. (See Bibliography to Chapter IV.)
Castle, W. E.
   1909. "A Mendelian View of Sex-heredity." *Science*,
   N. S., vol. 29, pp. 395–400.
Cuénot, L.
   1900. "Sur la détermination du sèxe chez les animaux."
   Bull. Sci. de la France et de la Belgique.

HEREDITY

MORGAN, T. H.
    1909. "A Biological and Cytological Study of Sex Determination in Phylloxerans and Aphids." *Journal of Experimental Zoölogy*, 7, pp. 239–352.
    1910. "Sex-limited Inheritance in Drosophila." *Science, N. S.*, 32, pp. 120–122.
    1911. "The Application of the Conception of Pure Lines to Sex-limited Inheritance and to Sexual Dimorphism." *The American Naturalist*, 45, pp. 65–78.

RUSSO, A.
    1909. "Studien über die Bestimmung des weiblichen Geschlectes." G. Fischer, Jena.

WILSON, E. B.
    1909. "Recent Researches on the Determination and Heredity of Sex." *Science, N. S.*, 29, pp. 53–70.
    1910. "The Chromosomes in Relation to the Determination of Sex." *Science Progress*, 5, pp. 570–592.

For references to the earlier literature see CUÉNOT and BATESON.

INDEX

INDEX

TWENTIETH CENTURY TEXT-BOOKS.

TEXT-BOOKS OF ZOOLOGY.

By DAVID STARR JORDAN, President of Leland Stanford Jr. University; VERNON LYMAN KELLOGG, Professor of Entomology; HAROLD HEATH, Assistant Professor of Invertebrate Zoology.

Evolution and Animal Life.

This is a popular discussion of the facts, processes, laws, and theories relating to the life and evolution of animals. The reader of it will have a very clear idea of the all-important theory of evolution as it has been developed and as it is held to-day by scientists. 8vo. Cloth, with about 300 illustrations, $2.50 net ; postage 20 cents additional.

Animal Studies.

A compact but complete treatment of elementary zoology, especially prepared for institutions of learning that prefer to find in a single book an ecological as well as morphological survey of the animal world. 12mo. Cloth, $1.25 net.

Animal Life.

An elementary account of animal ecology—that is, of the relations of animals to their surroundings. It treats of animals from the standpoint of the observer, and shows why the present conditions and habits of animal life are as we find them. 12mo. Cloth, $1.20 net.

Animal Forms.

This book deals in an elementary way with animal morphology. It describes the structure and life processes of animals, from the lowest creations to the highest and most complex. 12mo. Cloth, $1.10 net.

Animals.

This consists of "Animal Life" and "Animal Forms" bound in one volume. 12mo. Cloth, $1.80.

Animal Structures.

A laboratory guide in the teaching of elementary zoology. 12mo. Cloth, 50 cents net.

D. APPLETON AND COMPANY,
NEW YORK.   BOSTON.   CHICAGO.   LONDON.

Adolescence: Its Psychology and Its Relations to Physiology, Anthropology, Sociology, Sex, Crime, Religion, and Education.

By G. STANLEY HALL, Ph.D., LL.D.   Two vols., royal 8vo, gilt top.   Cloth, $7.50 net.

This work is the result of many years of study and teaching. It is the first attempt in any language to bring together all the best that has been ascertained about the critical period of life which begins with puberty in the early teens and ends with maturity in the middle twenties, and it is made by the one man whose experience and ability pre-eminently qualify him for such a task. The work includes a summary of the author's conclusions after twenty-five years of teaching and study upon some of the most important themes in Philosophy, Psychology, Religion, and Education.

The nature of the adolescent period is the best guide to education from the upper grades of the grammar school through the high school and college. Throughout, the statement of scientific facts is followed systematically by a consideration of their application to education, penology, and other phases of life.

Juvenile diseases and crime have each special chapters. The changes of each sense during this period are taken up. The study of normal psychic life is introduced by a chapter describing both typical and exceptional adolescents, drawn from biography, literature, lives of the saints, and other sources.

The practical applications of some of the conclusions of the scientific part are found in separate chapters on the education of girls, coeducation and its relations to marriage, fecundity, and family life, as seen by statistics in American colleges, with a sketch of an ideal education for girls.

Another chapter treats with some detail and criticism the various kinds and types of organization for adolescents from plays and games to the Y. M. C. A., Epworth League, and other associations devised for the young.

The problem of the High School, its chief topics and methods, is considered from the standpoint of adolescence, and some very important modifications are urged. It closes with the general consideration of the relations of a higher to a lower civilization from this standpoint.

D. APPLETON AND COMPANY, NEW YORK.

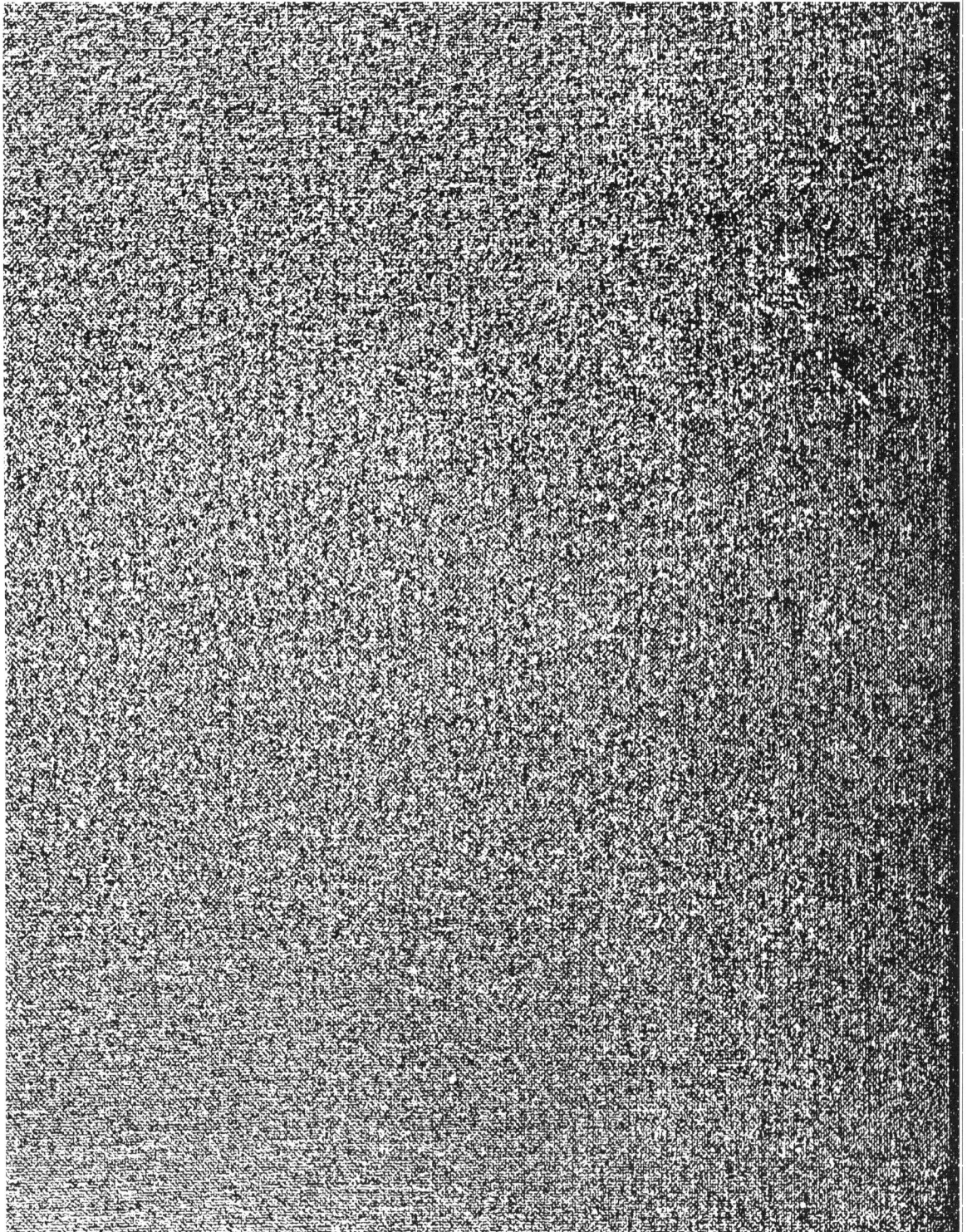

Printed in Great Britain
by Amazon

58540802R00137